Sustainable Design

Tomayess Issa · Pedro Isaias

Sustainable Design

HCI, Usability and Environmental Concerns

Second Edition

 Springer

Tomayess Issa
Curtin University School of Management
and Marketing
Perth, WA, Australia

Pedro Isaias 🄳
West Wing
UNSW Sydney
Sydney, NSW, Australia

ISBN 978-1-4471-7515-5   ISBN 978-1-4471-7513-1   (eBook)
https://doi.org/10.1007/978-1-4471-7513-1

Foreword

In order to help achieve the world's goals of addressing environmental issues, we are tasked with making our Information Technology (IT) systems and business processes more environmentally friendly and harnessing the potential of IT to solve or mitigate pressing environmental problems. Additionally, the IT sector is being scrutinized more closely owing to the effects that IT products and services have on sustainability throughout their lifecycle. To mitigate risks and maximize the advantages and possibilities offered by green IT, a comprehensive and strategy based on solid principles and best practices is required. Business executives, legislators, IT experts, academics, and students, as well as the general public, need realistic advice on how to utilize green IT. However, there are very few books on green IT that are both practical and helpful.

Many IT professionals, business personnel, and individuals who use IT are keen to explore and implement innovative ideas in this field. The combination of the principles of Human–Computer Interaction (HCI) with sustainability design is one such innovation. Previously, the primary goal of incorporating HCI in the design process was to create a user interface that was both efficient and effective in meeting the needs and requirements of prospective users. However, this book is intended to guide designers and users in the creation of IT systems and devices that meet future needs while incorporating HCI principles and sustainability in the design.Therefore, this work offers guidelines regarding the management and participation of users in the development of successful designs for various platforms such as websites, phones, tablets, and wearable devices. The intention of this book is to provide guidelines for the clean, minimalist design of systems that enable users to interact easily with the platform, while minimizing any negative impacts on the environment.

This book is for anyone interested in learning about sustainable design in human–computer interaction, usability, and environmental concerns in order to ensure a more sustainable ecosystem for coming generations. These designs have been evaluated by users in a number of countries comprising both developed and developing nations: Australia, Brazil, China, Germany, India, Norway, Singapore, South Korea, Sweden, the UK, and the USA. The data obtained from the research participants helped the researchers to determine the most important elements to include in the new sustainable design. Additionally, the writers will continue to include other subjects into to the textbook, including social and global problems, social networking, big data, and Internet of Things (IoT).

Finally, this book establishes the relationship between HCI, usability, and sustainable design and presents the most up-to-date knowledge on the aforementioned topics, as the majority of HCI authors are eager to develop frameworks, tools, techniques, and models that comply with sustainable design requirements.

Riyadh, Saudi Arabia   Abdulaziz Albahlal, Ph.D.
                                                      IT Professional, Consultant at Saudi
                                                       Authority for Data and Artificial
                                                           Intelligence (SDAIA)

Acknowledgements

We acknowledge the support of all people that contributed directly or indirectly to this book.

Tomayess Issa owe special gratitude first to God, the Almighty, and her parents: V. Rev. Fr. Boutros Touma Issa; and of course, to the Spirit of Bathqyomo Marine Khoury-Issa (2016+). Her sister Dr. Theodora, her brother Dr. Touma, his wife Siba, and their daughter Talitha; her Sister Tamara, her husband Tony, and their children Tabitha, Antoinette, Jacob, and finally to her bother Theodore, his wife Mary, and their children Cephas and Mary; for their continuous support, and encouragement since without their help, this work would never have been completed. Tomayess would like to acknowledge the support of The School of Management and Marketing at the Faculty of Business and Law of Curtin University, Australia, for their inspiration and encouragement.

Pedro Isaias would like to acknowledge The School of Information Systems and Technology Management (SISTM) from the University of New South Wales, Sydney, for their support. Also, a special acknowledgment to Sara Pifano from ISRLab for her continuous support. Finally, he would like to acknowledge the support of his family. Without their assistance, this work would not have been possible at all.

Perth, WA, Australia   Tomayess Issa
Sydney, NSW, Australia   Pedro Isaias

Contents

Chapter 1
Introduction

Abstract In this book, we examine the importance of human–computer interaction (HCI), usability, and sustainability, as well as the concept of sustainable design, in the information communication and technology (ICT) sector. ICT usage by businesses and individuals has become a significant instrument for searching, conducting research, communication, entertainment, commerce, and information. However, the recycling of ICT hardware is becoming a major dilemma for businesses and individuals since it is not simply a matter of concern for environmental damage or a solution to an environmental problem. Designers, businesses, and individuals must make a concerted and collaborative effort to tackle environmental concerns by developing new ICTs that incorporate sustainable designs in order to meet the current and future needs of businesses and individuals while minimizing the negative effects on the environment. This book discusses the features of sustainable design and proposes a New Participative Methodology for Sustainable Design (NPMSD). Finally, the objectives of this book, which are directly aligned with the UN's Sustainable Development Goals 7, 9, 12 and 13, are to create an innovative sustainable technology that facilitates recycling, waste reduction, the conservation of raw material for the seventh generation, and the use of clean and efficient energy.

1.1 Introduction

Computer technology, Internet technology, and associated systems are essential tools in the twenty-first century since businesses and individuals have come to depend increasingly on them rather than on the traditional systems used to achieve the same ends. Today's technology is more capable of managing and assisting businesses and individuals to complete their tasks far more efficiently. Not only is there a proliferation of stand-alone computers; networking on a global scale has increased enormously as a result of the Internet, World Wide Web, social networks, mobile systems, intelligent environments, and other technological developments. The increasing usage of ICT worldwide presents a new challenge to HCI researchers and practitioners who need to cater for the needs of businesses and individuals while ensuring that emerging technologies are more sustainable for both current and future needs. HCI is the study

© Springer-Verlag London Ltd., part of Springer Nature 2022   1
T. Issa and P. Isaias, *Sustainable Design*,
https://doi.org/10.1007/978-1-4471-7513-1_1

of the interaction between humans and complex technology and concerns the way that the current input and output of technologies influence the interaction between a user and the interface. HCI draws on many disciplines, but it is in "computer science and systems design that it must be accepted as a central concern, and HCI involves the design, implementation, and evaluation of interactive systems in the context of the user's task and work" (Dix et al. 1993, p. 4).

Therefore, HCI researchers should consider not only productivity and customer satisfaction, but also human factors that affect the "acquisition, disposal, renewal, and reuse and design for sustainability" (Dillahunt et al. 2010, p. 1). In addition, they should attempt to create and develop technologies which are more effective and efficient and should study the "social and communal aspects of technology use and effective and aesthetic aspects of design" (Sengers et al. 2006, p. 1683). To achieve this, they must consider the different perspectives of users and designers in order to understand their notions of design, attitudes, ethnography, and user empathy and seek to develop new technologies that address sustainability goals for the current and future generations (Busse et al. 2009; Sengers et al. 2009a).

Hence, HCI researchers, businesses, and individuals should incorporate the concept of "green" technologies in their designs, since the adverse effects of current technologies are causing major problems for the environment. In addition, a new system design should integrate sustainability principles to ensure that it is more sustainable, user-friendly, safe, efficient, and effective for use by businesses and individuals. This can be achieved by studying and understanding potential users' desires and requirements. Furthermore, this book will examine the importance of HCI, usability, and sustainability in respect to design systems, thereby raising the awareness of HCI practitioners and academics regarding the development of new technologies, bearing in mind the future generations. In addition, a new sustainable design model will be developed to promote the notions of HCI, usability, and sustainability when developing new devices now and in future.

This book is organized as follows: Introduction, HCI, usability, user participation in the system development process, physical, cognitive affective engineering, Color, prototyping and navigation, guidelines and principles design, evaluation and testing; task analysis, models, and methodologies and the New Participative Methodology for Marketing Websites (NPMMW), the New Participative Methodology for Sustainable Design (NPMSD), and future ICT.

1.2   Human–Computer Interaction

The term "human–computer interaction" was adopted in the mid-1980s to describe this new field of study. HCI "is a discipline concerned with the design, evaluation, and implementation of interactive computing systems for human use and with the study of major phenomena surrounding them" (Preece et al. 1994, p. 7). However, this field is now "concerned with understanding, designing for, and evaluating a wider range of user experience aspects" (Sharp et al. 2011, p. 18). Therefore, the reason for studying HCI in the development process is to create interactive computer systems that are usable as well as practical (Head 1999).

HCI relates to several stages in the development process, including the design, implementation, and evaluation of interactive systems, in the "context of the user's task and work" (Dix et al. 2004, p. 4). According to Vora (1998), HCI implementation requires a massive range of skills, including an understanding of the potential users, their tasks, and environments, software engineering capabilities, and graphical interface.

Because designers often have a poor understanding of HCI issues, they need to know the needs and values of prospective users and the tasks that the technology is required to support. Most importantly, they need to be able to translate that knowledge into an executable system. This can be accomplished by establishing a good interface design that enables the user to interact with and navigate the interface with ease; ultimately, the interface should give the user more control of the site.

The main purpose of incorporating in the design the features related to HCI is to develop an efficient and effective user interface to suit user requirements and wishes. Hence, HCI specialists need to involve the users in the design process, integrating different kinds of knowledge and expertise and making the design process iterative (Preece et al. 1994). It was noted that HCI design should: be user-centric, integrate knowledge from different disciplines, be highly iterative, and include an effective usability evaluation. This process enables feedback to be obtained regarding the negative and positive aspects of prototypes. It is important that the way in which people interact with computers is intuitive and clear. However, the designing of a system that contains appropriate HCI features is not always straightforward, as evidenced in the many poorly designed computer systems. One of the challenges of HCI design is to keep abreast of technological developments and to ensure that these are harnessed for maximum human benefit.

The goals of HCI are to produce usable, safe, and functional systems. These goals can be summarized as safety, utility, effectiveness, efficiency, and appeal. These goals relate to the services that the system provides and how quickly the tasks can be achieved, and ensuring that users like the system. Moreover, Haklay (2010, p. 5) indicated that HCI aims to create systems which provide functionality to meet the needs of businesses and individuals. In addition, in order to develop or improve their design, HCI specialists should understand how system design can support users in an effective and efficient manner, and how users intend to use computers systems. Finally, Bodker, Byrne, and Boye (cited in Maceli and Atwood 2011) describe the three waves of HCI: humans as factors, actors, and crafters. Therefore, all information interfaces including websites should ensure a good interaction with users and vice-versa, by providing efficiency and safety, and making them more enjoyable for users.

1.3   Usability

Usability refers to the "quality of the interaction in terms of parameters such as time taken to perform tasks, number of errors made, and the time to become a competent user" (Benyon et al. 2005, p. 52). Also, usability "is a quality attribute that assesses how easy user interfaces are to use. The word "usability" also refers to methods for improving ease-of-use during the design process" (Nielsen 2003). Furthermore, Shackel (2009, p. 340) states that usability is the "capability in human functional terms to be used easily and effectively by the specified range of users, given specified training and user support, to fulfill the specified range of tasks, within the specified range of environmental scenarios."

By including a usability evaluation stage, a software development team can determine the positive and negative aspects of its prototype releases and can make the required changes before the system is delivered to the target users. Usability evaluation involves observing the users to "see what can be improved, what new products can be developed" (McGovern 2003), and is "based on human psychology and user research" (Rhodes 2000). HCI specialists observe the participants who are required to execute a real task on a site, and these observations will allow them to obtain a comprehensive insight into the way the participants have experienced the site.

From the user's perspective, usability is a very important aspect of the development process as it can mean the difference between "performing a task accurately and completely or not" and the user "enjoying the process or being frustrated" (Usability First 2002). Moreover, if usability is not an integral part of user interface design, then users will become very frustrated with the interface. In general, usability is an essential HCI concept and is concerned with making systems easy to learn, easy to use, and with minimal error frequency and severity. In order to develop a successful system with good usability, HCI specialists need to understand and include various factors in their designs, namely organizational, social, and psychological factors that determine the extent to which people effectively operate and make use of computer technology. They need to develop tools and techniques to help designers ensure that computer systems are suitable for the activities for which people will use them, and achieve efficient, effective, and safe interaction in terms of both individual and group computer interactions. These factors should be considered very carefully during the design stage, as most of the users should not have to change radically to "fit in" with the system; rather, the system should be designed from the outset to meet their requirements (Preece et al. 1994).

Furthermore, Sharp et al. (2011) confirm that usability goals should be considered by designers and HCI specialists to ensure that the user interface is easy to learn and remember, effective and efficient to use, and with fewer errors and good utility. These goals can be applied to the design of an interactive system to ensure its usability. These principles guide the developers during the system design stage. Together with the aforementioned principles, an important additional key factor is "utility" which is the functionality enabling users to "do what they need or want to do" (Preece et al. 2002, p. 16). In other words, system designers need to keep this question in mind:

"does it do what users need?" (Nielsen 2003). Hence, usability and utility are equally important in the development process and must be integrated.

To summarize, it was noted that HCI and usability are essential factors to consider when designing and developing a user interface which is more efficient and effective and produces user satisfaction rather than frustration. In order for the interface to have these attributes, potential users should participate in the design from the outset. Folstad et al. (2010) and Issa et al. (2010) reiterate that user participation is essential in the system development process and users should be present during this process to share their opinions, especially from the initial planning stage through to the maintenance stages and procedures.

Furthermore, according to Issa et al. (2010), user participation in the system development process will prevent user frustration, thereby reducing training time, and will ensure that the system design is aligned with users' requirements. Finally, Nies and Pelayo (2010) confirm the necessity of involving users in the system development process so that the design meets their requirements.

1.4 Sustainability

Before discussing the term "sustainable design", we need to discuss the notion of "sustainability", since these two concepts are related in terms of benefitting human and natural resources that will be needed in the future (Weybrecht 2010). The term "sustainability" is derived from the Latin word *sustinere*, which means "to hold up", "sustain", "give support to", "to bear", or to "keep to" (Sadri and Goveas 2013; Youmatter 2021). Gro Harlem Brundtland from the World Commission on Environment and Development first coined the term "sustainability" in 1983. The term "sustainable development" was first referred to in 1987 in the Brundtland Report on "Our Common Future", where Brundtland's report urged businesses and individuals to progress toward economic development in a way that could be sustained without destroying the natural resources or the environment for the next generation (Newton 2003).

Erek et al. (2009, p. 2) define sustainability as "a survival assurance meaning that an economical, ecological or social system should be preserved for future generations and, thus, necessary resources should only be exploited to a degree where it is possible to restore them within a regeneration cycle." This suggests that businesses and individuals must protect the current infrastructure so that resources are available for the next generation. The notion of sustainability is highly significant in the twenty-first century since, increasingly, businesses and individuals are now required to think in terms of delivering "solutions rather than products and seek to define their markets in terms of customer activities and outcomes rather than products and services" (Jeffers 2009, p. 263).

The integration of sustainability in businesses and in individuals' strategies will be highly advantageous in terms of cost reduction, resources preservation, conformity to legislation, improvement of reputation, maintaining happier customers and stakeholders, attracting capital investment, and capitalizing on new opportunities (Nidumolu et al. 2009; Prahalad and Rangaswami 2009; Sharma et al. 2010; Smith and Sharicz 2011). Finally, Kendall and Kendall (2010) indicated that sustainability will assist businesses, stakeholders, individuals, and society in general.

The integration of sustainability in business strategies should be aligned with the project needs and business proposals of a particular division or even the whole company. According to Weybrecht (2010), the adoption and application of sustainability in businesses will achieve the following advantages: cost reduction; preservation/saving of resources; compliance with legislation; enhanced reputation that differentiates businesses; securing quality employees; satisfy customer needs; meeting of stakeholder expectations; attracting of capital investment; and capitalizing on new opportunities.

These advantages will make the business unique in the market locally and globally, since sustainability is already a part of how business is done by many companies worldwide. However, the nature of the business is not as important as its ability to continue. Currently, despite the high cost of integrating sustainability in business strategy, this is outweighed by the long-term advantages to businesses and to society at large. However, the full benefits of sustainability cannot be achieved unless it is embedded in a firm's culture and business strategy, and unless it has the full cooperation of the CEO.

To integrate the sustainability factor in the business strategy, the project manager should collect all the necessary information about what is happening in his/her company at all levels of the business hierarchy. Once the required information has been collected, it is necessary to secure everyone's cooperation so that management and all employees have the same positive attitude toward sustainability. The project manager must pick the correct moment to disseminate the notion of sustainability adoption throughout the organization. The advantages and disadvantages of integrating sustainability in a business strategy should be put to management whose role it is to inform staff of any changes that this requires. Furthermore, in terms of sound business practice, the various attitudes of staff, together with their roles, backgrounds, and personalities, should be taken into account.

Moreover, the project manager must make a strong case by outlining the benefits of a sustainability policy and the disadvantages if the organization does not address this issue. This adoption of sustainability in the organization structure will be useful when hiring new staff.

On the other hand, the integration of sustainability into an organization's business strategy can bring some risks associated with governance failure, cost, fraud, scandals, increase insurance due to environmental disasters, compliance breaches, increased competition for and cost of raw materials, security, and privacy issues. (Jeffers 2009; Pichlak and Szromek 2021; Xiao et al. 2013) These potential issues may cause some firms to reject the integration of sustainability in their business strategy. Therefore, there is a need for awareness, collaboration, and communication, between academics, researchers, and organizations to minimize these disadvantages in the market and to start integrating sustainability in their strategies and operations to minimize the negative impact that these have on the natural environment.

Sustainability as an integral part of an organization's strategy requires under-standing, consideration, and tolerance at all levels of the organization as well from its stakeholders. The strategy should be easy, straightforward, dynamic, and easy to implement. Finally, patience must be exercised when changing the mindset and attitudes of staff and stakeholders prior to introducing sustainability-related strategies.

According to Moscardo et al. (2013), sustainability requires a long-term orienta-tion and commitment to changing the way businesses conduct their activities in order to balance current needs with those of future generations. Furthermore, there should be recognition that business is part of a complex system comprising environmental, social, and economic activities. As part of the business strategy, training should be provided to staff to increase their awareness of and knowledge about sustainability.

This learning should not be limited to staff; specific training should be available to stakeholders and the community to make them aware of all the issues concerning sustainability, since the needs of the business should and must match the needs of stakeholders, society, the economy, and the environment. The integration of sustain-ability in business strategy will increase a company's revenue and improve its repu-tation in the market locally and globally. Finally, sustainability is a complex area that is continually evolving. This means that both management and employees should be made aware of changes through training and keeping up to data with the latest developments in this field. All firm personnel should be aware of the benefits and risks associated with the implementation of sustainable practices, the issue of green washing, and the need to change the mindset of individuals and the company culture.

Finally, sustainability is considered an incorporated design to environmental, social, and economic concerns, which take the lead to a long-term sustainable profit growth (Pedol et al. 2021; Portney 2015; Robertson 2021; Scoones 2007; Thiele 2016) (see Fig. 1.1).

Figure 1.1 depicts the ideal balance that should exist between the three pillars of sustainability representing the triple bottom line concept that many industries have adopted to ensure that they remain financially viable while safeguarding society and the environment.

1.5 Sustainable Design

According to Nathan et al. (2008), the terms "sustainable" and "green" are used ubiq-uitously within businesses and by individuals locally and globally. Currently, these notions play a major role in business and individual strategies; therefore, any design should ensure that whatever is created and developed should first meet the current users' and businesses' requirements and, of course, those of the next generation.

Stelzer (2006, p. 4) defines sustainable design as the "fundamentally a subset of good design. The description of good design will eventually include criteria for the creation of a healthy environment and energy efficiency." Silberman and Tomlinson (2010, p. 3470) discuss the relationship between sustainable design and

Fig. 1.1 Sustainability
Summary – Prepared by
Authors

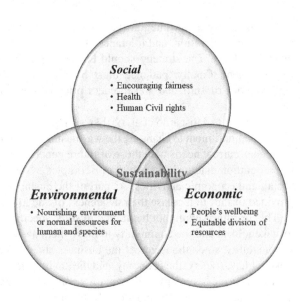

HCI, confirming that previously HCI researchers were concerned with "What do users do? When? How often? Why? How do they feel about it? What do know about what they are doing? How do they know?".

Nowadays, however, HCI researchers should understand the relationship between users and technologies, and how this can assist designers to simplify more sustainable user practices. Moreover, DiSalvo et al. (2010) confirm that HCI researchers and top management should be encouraged to collaborate in the design and development of applications, interfaces, equipment, and services with more sustainable effects; in addition, this design should comply with the principles of economic, social, and ecological sustainability.

Smith and Sharicz (2011) posit that HCI and information technology researchers and professionals must take into consideration the environmental impact of the design of current and future technologies, so that practitioners are aware of the environmental impact of the technologies they use. Most importantly, sustainable design should meet users' needs. Sustainable design can be achieved only with the awareness and innovation of designers and users. Awareness can produce opportunities to be unique and exceptional in design, and this can lead to creativity and innovations in research.

Awareness of the need for change can contribute to the improvement of the environment, to social equity, and to growth and profit in the expanding global community. This awareness will lead the designers to action orientation, learning, and excitement, and to a new level of caring based upon new knowledge and commitment. To achieve this, participation in sustainable design is essential, and designers must take into account the opinions and perspectives of potential users and invite them to assist with the design, since designers should not act independently.

Currently, the world is under pressure from human actions that threaten sustainability. According to Reuters (2020), at the global level, the quantity of e-waste

Fig. 1.2 E-waste generation per continent in 2019 Adopted by (Forti et al. 2020)—Prepared by the authors

generation in 2020 was around 53.9 million tones. This number will be increased to 74.7 by 2030 (Tiseo 2021), representing an increase in growth rate of 4% to 5% if developers maintain their current practice when designing, without integrating sustainability in their system design. Figure 1.2 shows the e-waste generated per continent; Asia generated 24.9 MT in 2019, while Oceania generated only 0.7 MT in 2019 (Forti et al. 2020).

As shown by the statistics in Fig. 1.2, the world is sustaining a considerable amount of negative environmental impact due to unsustainable human activity, necessitating a plan of action to change the way we live. Therefore, designers, users, and organizations should focus on and commit to designing products and electronic devices that comply with the principles of social, economic, and ecological sustainability (Rahm-Skågeby and Rahm 2021; Soden et al. 2021).

In this book, a sustainable design step is added to the design stage, since this step will allow designers to consider the factors necessary for the development of new smart technology and portable devices that incorporate sustainability. This step applies to design that takes these issues into account: manufacturing, energy consumption, recycling, safety, efficiency, and social impact. Amalgamating, incorporating, and integrating the sustainable step in the new portable devices will minimize current environmental problems such as carbon emissions, raw materials usage, and global warming. It is imperative that these problems be addressed immediately before further irreparable damage is done to the planet.

Furthermore, to understand the sustainable design concept, in this book awareness, advantages and disadvantages of sustainable design will be examined. Eleven countries, comprising both developing and developed nations, will be examined via a survey to determine whether the concept of sustainability must be an important consideration of users and companies now and in the future.

Figure 1.3 presents a set of factors which need to be considered in regard to sustainable design awareness, advantages, and disadvantages. For the sustainable design awareness, several studies (Atz et al. 2021; Issa et al. 2020; Jaffe et al. 2020;

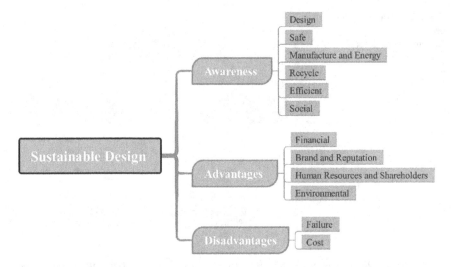

Fig. 1.3 Sustainable design: awareness, advantages and disadvantages—Prepared by Tomayess Issa

Petrova et al. 2019; Ramani 2010; Robertson 2021; Sikdar et al. 2017) indicate that designers and HCI experts must develop and design new portable devices that include a sustainable factor in their design strategy, since the new devices must be easy to upgrade and recycle, and should reduce the carbon footprint, air pollution, climate global warming, and over-consumption and waste of resources.

Furthermore, the production of new devices should require less energy and raw materials, since they will be manufactured from recycled, recyclable, and renewable materials. Finally, the new devices should comply with good ethical principles.

Regarding the advantages of a sustainable design, new devices should be able to minimize the consumption of energy and raw materials, be "greener", improve CSR and TBL, increase vendor productivity, and reduce carbon footprint, emission, pollution, and health hazards.

While the possibility of failure and the issue of cost are two of the disadvantages of sustainable design, there are several other associated risks (listed on p. 7). These disadvantages should be tackled by communication, collaboration, cooperation, and connection between designers, HCI experts, organization's decision makers in order to reap the benefits of sustainable design and minimize the disadvantages. Therefore, sustainable design aims to design devices and services that apply environmental sustainability criteria.

Finally, sustainable design is one means by which we can make our world better. However, in order to achieve this, we need to have the right motivation, awareness, knowledge, commitment, trust, and loyalty. Firms need to act quickly and to incorporate sustainability in their business strategy and in the design of systems that are environmentally friendly and conserve resources for the next generation.

1.6 Methodology

To address the aforementioned objectives of this book, an online survey was conducted to examine users' attitudes toward sustainability and sustainable user interface design in Australia, Brazil, China, Germany, India, Norway, Singapore, South Korea, Sweden, UK, and USA. The survey data assisted the authors to discover any new factors that should be incorporated in the new sustainable design model. The design of the online survey questionnaire was based on the findings of the literature. It comprised three sections: background; sustainable design, and advantages and disadvantages of sustainability. The questionnaire enabled users to suggest new factors for the new sustainable model and confirmed the theoretical significance of this book.

The online survey was considered appropriate for this study as it offers anonymity, is less expensive, and is more accessible (Braun et al. 2020; Issa 2013; Kocher 2015; Miller et al. 2020). However, technical failure, computer viruses, Internet crimes, hacking, and privacy are considered the disadvantages of online surveys, and these factors can reduce the response rate (Fan and Yan 2010; Smyk et al. 2021).

1.7 The Initial Sustainable Step in the New Participative Methodology for Sustainable Design

Sustainability is now generally accepted by most organizations as an important part of corporate citizenship. The concept of sustainability is based on the notion that our actions should not cause irreparable harm to our social and environmental infrastructure. It calls for us to take responsibility for our actions and ensure that we act to improve or change our current way of living to avert social, environmental, and ecological crises.

In order for this to occur, and deriving from Dyllick and Hockerts (2002) and McDonough and Braungart (2002) models of corporate sustainability, Young and Tilley (2006) proposed an integrated model of corporate sustainability which links together six criteria that a sustainable business will need to satisfy. The six criteria are: (1) eco-efficiency, (2) socio-efficiency, (3) eco-effectiveness, (4) socio-effectiveness, (5) sufficiency, and (6) ecological equity. However, further theoretical development is required in order to create an effective, integrated approach to applying the six criteria.

Erek et al. (2009, p. 2) stated, "Sustainability has been extensively discussed within corporate management under the synonyms of corporate social responsibility (CSR), greening the business eco-efficiency or eco-advantage." To ensure that organizations develop and adhere to a sustainable development strategy, management should consider aspects of value creation that would benefit its employees, users, and stakeholders by encouraging all participants to be environmentally and socially responsible corporate citizens.

In line with the integration of a sustainability strategy into technology, various studies on HCI, usability, and sustainability were examined to study the impacts of ICT on the environment (Bevan 2001; Bodker 2006; Dillahunt et al. 2010; DiSalvo et al. 2010; Gerlach and Kuo 1991; Mann 2009; Nathan et al. 2008; Ramani 2010; Sengers et al. 2006, 2009b; Silberman and Tomlinson 2010; Te'eni et al. 2007; Wilson and Borras 1998). It was noted that the recycling of ICT products is becoming a major dilemma for businesses and individuals, since it is not simply a matter of concern for environmental damage or a solution to an environmental problem.

Designers, businesses, and individuals must collaborate in making a concerted effort to tackle the environmental concerns by developing new ICTs that incorporate sustainable design while meeting the current and future needs of businesses and individuals. Therefore, in this book, we propose and discuss a New Participative Methodology for Sustainable Design for smart new technology and portable devices. A review of the current literature (Comm and Mathaisel 2015; Gauthier 2015; Issa 2014; Issa and Isaias 2014; Kemp 2015; Nidumolu et al. 2009; Pan et al. 2015; Shaw et al. 2015; Stapledon et al. 2015; Stelzer 2006; Wals 2014; Wang et al. 2015a, b) yielded the initial factors for the sustainable step which includes design, safety, manufacture and energy, recycling efficiency, and social impact (see Fig. 1.4).

These critical factors are included in the first draft of the New Participative Methodology for Sustainable Design. The authors added new characteristics and critical factors to the new sustainable model during the design stage under the.

Fig. 1.4 Initial factors for the sustainable step—Prepared by Tomayess Issa

Fig. 1.5 New participative methodology for marketing websites' (NPMMW)—Prepared by Tomayess Issa

New Participative Methodology for Marketing Websites' (NPMMW) (see Fig. 1.5). The NPMMW includes all the necessary stages and steps which are required to develop an efficient and effective device that meets users' needs.

Figure 1.6 illustrates the first draft of the new Sustainable Model, which will be part of the design stage under the NPMMW methodology. This model will use all the stages and steps, which belong to the NPMMW model to ensure that the new devices meet users' requirements and needs.

According to Stelzer (2006), sustainability is primarily a subset of design. Design is an exercise in meeting the challenges inherent in any situation that requires improvement or mediation. Ultimately, any design solution will need to create products and environments for a living earth with limited resources. The criteria for successful design will lead to the creation of a healthy present and a prosperous future; thus, by extension, the attainment of sustainability is a question of good design.

1.8 Outline of the Book

This book comprises eight chapters, each of which presents and describes the concepts and approaches, which will give readers the necessary information required to understand the study's proposals and their significance. The chapters' topics have

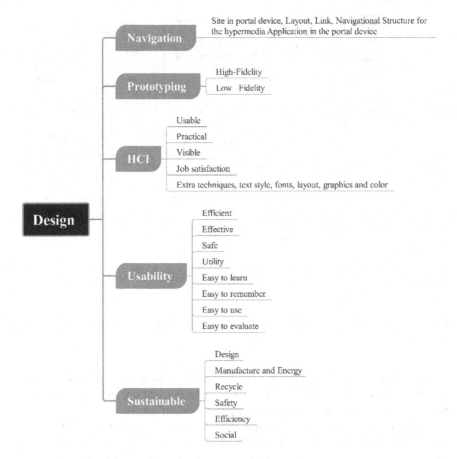

Fig. 1.6 Sustainable step in the new participative methodology for sustainable design—Prepared by Tomayess Issa

been carefully selected by the authors to ensure that readers will learn and put into practice the skills which are required to develop an efficient, effective, user-friendly, and sustainable design. From Fig. 1.7, it is noted that readers will learn the definition and the importance of HCI and usability in respect to user satisfaction and the efficiency, effectiveness, and user friendliness of the system (Lee and Koubek 2010; Nicolson et al. 2011).

In this chapter, the authors discuss in detail the notions of human–computer interaction and usability and make a case for the relationship between sustainability and HCI, since this issue is becoming essential in the system development process. Although designers should integrate sustainability in their design and framework, innovative designs should not only include functions that satisfy the consumers, but should also be sustainable (Ramani 2010). Therefore, designers, users, and top

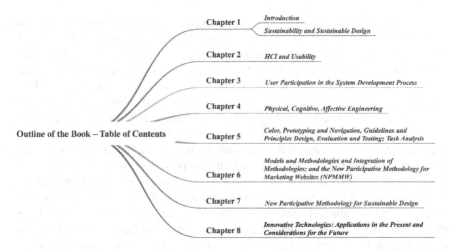

Fig. 1.7 Outline of the book—table of contents—Prepared by the Authors

management must, together, work smarter and harder and more creatively if "we are going to help save our planet from ourselves" (Ramani 2010, p. 1).

Several scholars (Grover et al. 2020; Jaffe et al. 2020; Petrova et al. 2019; Sharma et al. 2010; Thatchenkery et al. 2010) propose that an agreement should be developed between designers, users, and all stakeholders, encouraging them to work collaboratively on a sustainable design for devices such as smart phone, laptops, and other portable devices to meet the needs of current and future generations while minimizing the negative environmental impacts.

Chapter 2 encourages readers to become familiar with the principles and guidelines for developing a system that ensures successful human–computer interaction and usability.

Chapter 3 focuses on user participation. In the first stage in particular, and throughout the system development process, users should be invited to express their opinions about the design in order to prevent potential user frustration.

Chapter 4 examines the differences between physical, cognitive, and affective engineering, since these topics will assist readers to understand that design is not limited to layout, navigation, and color, but that other aspects should also be taken into consideration in the design process. These topics will stimulate discussion about the interaction and relationship between human and machine, ergonomics, and development concerns such as memory, attention span of users, and reduction of complexity between the goals of cognitive engineering, speed and accuracy and finally, effectiveness, i.e., making the interface more appealing, entertaining, enjoyable, engaging, and fun (Te'eni et al. 2007).

Chapter 5 discusses the importance of color, navigation, and prototyping in the system development process (Bonnardel et al. 2011; Cyr et al. 2010), as designers and users should be satisfied with the final sketches before coding and implementation occur. Furthermore, the authors will discuss the significance of evaluating and testing following every stage of the system development process. In respect to the evaluation, the authors will address the issue of why and what and when to evaluate the output of the system development process; they will also discuss the difference between formative and summative evaluation. Additionally, the testing concept will be discussed in this section to distinguish between evaluation and testing and their place in the system development process (Issa et al. 2010; Petre et al. 2006). To assess and evaluate an interface (including the website), readers should understand the concept underlying the design principles and guidelines which will be introduced in this chapter. A knowledge of design principles is essential since readers will learn how to evaluate interfaces (including the websites) professionally and from different perspectives in terms of promotion of trust, diversity of users, affordability and performance, matching information representations needed with that presented, designing for errors, and providing, enjoyable, and satisfying interaction. On the other hand, the design guidelines will assist readers to evaluate and assess the interface (including the website) in terms of control and feedback, direct manipulation, metaphor, consistency and aesthetic appeal (Preece et al. 1994, 2002; Te'eni et al. 2007). Finally, readers will learn three aspects of task analysis: Task, Action and Goals (Galitz 2007; Shneiderman and Plaisant 2010). These concepts are very important in the design process since they assist both designers and users to identify the tasks which must be accomplished in order required in order to achieve specific goals.

To ensure that interfaces are developed successfully and do not cause frustration for users, Chap. 6 introduces a series of methodologies to demonstrate the sequential stages and steps, which are required to develop a system. The chapter explains the activities, techniques, and tools which are used for developing user-friendly interfaces.

Chapter 7 discusses the New Participative Methodology for Sustainable Design and identifies the new factors that must be considered when developing a sustainable design for today and for the future via a new survey which was employed in Australia, Brazil, China, Germany, India, Norway, Singapore, South Korea, Sweden, UK, and the USA.

Furthermore, Chap. 8 touches on other topics such as social, global issues and social networking, big data, IoT (Internet of Things), and other topics. The evolution of information and communication technologies is occurring at an unprecedented rate, with repercussions on all areas of society. Developments in key innovative technologies are at the center of revolutionary advancements in education, business, health, and government. At the same time, technological progress poses diverse challenges that can compromise inclusivity, sustainability, and it can increase security, ethical and privacy concerns. This chapter examines the key technological trends associated with social networks, artificial intelligence, IoT and big data, and extended reality and their application and challenges in several sectors. It also considers the fundamental role that digital divide and sustainability play in a scenario of rapid technological development.

Finally, this book establishes and consolidates the relationship between HCI, usability, and sustainable design and provides the latest information in respect to the previous topics, since the majority of HCI researchers are keen to develop frameworks, tools, techniques, and models to meet the sustainable design requirements.

1.9 Conclusion

In this chapter, we discussed the concepts related to the features of sustainable design. To establish the new sustainable model, an initial model was created, and an online survey was conducted in Australia to determine users' attitudes to sustainability and a sustainable user interface design. The online survey results are discussed in Chap. 7; later, we identify the new factors to be incorporated into the final sustainable model.

References

Atz U, Van Holt T, Douglas E, Whelan T (2021) The return on sustainability investment (ROSI): monetizing financial benefits of sustainability actions in companies. In: Sustainable consumption and production, vol II. Springer, pp 303–354

Benyon D, Turner P, Turner S (2005) Designing interactive systems: a comprehensive guide to HCI and interaction design, 2nd edn. Pearson Education Limited, UK

Bevan N (2001) International standards for HCI and usability. Int J Hum Comput Stud 55(4):533–552. https://doi.org/10.1006/ijhc.2001.0483

Bodker S (2006) When second wave HCI meets third wave challenges. In: Proceedings of the 4th nordic conference on human computer interaction: changing roles, Oslo, Norway

Bonnardel N, Piolat A, Le Bigot L (2011) The impact of colour on website appeal and users' cognitive processes. Displays 32(2):69–80. https://doi.org/10.1016/j.displa.2010.12.002

Braun V, Clarke V, Boulton E, Davey L, McEvoy C (2020) The online survey as a qualitative research tool. Int J Soc Res Methodol 1–14

Busse D, Blevis E, Howard C, Dalal B, Fore D, Lee L (2009) Designing for a sustainable future. In: Proceeding of the seventh ACM conference on creativity and cognition, Berkeley, California, USA, pp 493–494

Comm CL, Mathaisel DF (2015) Designing an engineering system for sustainability. Applied mechanics and materials. Trans Tech Publ, pp 474–478

Cyr D, Head M, Larios H (2010) Colour appeal in website design within and across cultures: a multi-method evaluation. Int J Hum Comput Stud 68:1–21

Dillahunt T, Mankoff J, Forlizzi J (2010) A proposed framework for assessing environmental sustainability in the HCI community. In: CHI 2010, Atlanta, GA, USA, pp 1–3

DiSalvo C, Sengers P, Hronn Brynjarsdottir P (2010) Mapping the landscape of sustainable HCI. In: CHI 2010, Atlanta, GA, USA, pp 1975–1984

Dix A, Finlay J, Abowd G, Beale R (1993) Human computer interaction. Pearson Prentice, Hall England

Dix A, Finlay J, Abowd G, Beale R (2004) Human-computer interaction, 3rd edn. Pearson Education Limited, England

Dyllick T, Hockerts K (2002) Beyond the business case for corporate sustainability. Bus Strateg Environ 11(2):130–141

Erek K, Schmidt N-H, Zarnekow R, Kolbe LM (2009) Sustainability in information systems: assortment of current practices in IS organizations. In: Americas conference on information systems (AMCIS), San Francisco, pp 1–9

Fan W, Yan Z (2010) Factors affecting response rates of the web survey: a systematic review. Comput Hum Behav 26(2):132–139

Folstad A, Anda B, Sjoberg D (2010) The usability inspection performance of work-domain experts: an empirical study. In: Interacting with computers, pp 75–87

Forti V, Baldé CP, Kuehr R, Bel G (2020) The global e-waste monitor 2020: quantities, flows, and the circular economy potential. https://www.google.com/url?sa=t&rct=j&q=&esrc=s& source=web&cd=&ved=2ahUKEwj7hdmR4aPzAhWaf30KHf9ADaoQFnoECAMQAQ&url= https%3A%2F%2Fwww.itu.int%2Fmyitu%2F-%2Fmedia%2FPublications%2F2020-Publicati ons%2FEN-Global---E-waste-Monitor-2020.pdf&usg=AOvVaw1j18vfwo-HmeqNs-0ypoBw. Accessed 26 Sept 2021

Galitz W (2007) The essential guide to user interface design an introduction to GUI design principles and techniques. Wiley, USA

Gauthier G (2015) A usability evaluation of a website focusing on the three initial steps of the conflict/resolution process for union members. http://scholarspace.manoa.hawaii.edu/bit stream/handle/10125/35855/Gauthier_Final_Paper_Scholarspace.pdf?sequence=1. Accessed 29 Sept 2021

Gerlach JH, Kuo F-Y (1991) Understanding human computer interaction for information systems design. MIS Q 14(4):526–549

Grover R, Emmitt S, Copping A (2020) Critical learning for sustainable architecture: opportunities for design studio pedagogy. Sustain Cities Soc 53:101876

Haklay M (2010) Interaction with geospatial technologies. Wiley

Head AJ (1999) Design wise. Thomas H Hogan Sr.

Issa T (2013) Online survey: best practice. In: Isaias P, Nunes MB (eds) Information systems research and exploring social artifacts: approaches and methodologies. IGI Global, USA, pp 1–19. https:// doi.org/10.4018/978-1-4666-2491-7.ch001

Issa T (2014) Online shopping and human factors e-commerce platform acceptance: suppliers, retailers, and consumers. Springer, UK

Issa T, Isaias P (2014) Promoting human-computer interaction and usability guidelines and principles through reflective journal assessment. In: Emerging research and trends in interactivity and the human-computer interface. IGI Global, pp 375–394. https://doi.org/10.4018/978-1-4666-4623-0.ch019

Issa T, Issa T, Issa TB, Isaias P (2020) Sustainability awareness and green information technologies. Springer

Issa T, Turk A, West M (2010) Development and evaluation of a methodology for developing marketing websites. In: Martako D, Kouroupetroglou G, Papadopoulou P (eds) Integrating usability engineering for designing the web experience: methodologies and principles. IGI Global Publishing, pp 103–123

Jaffe SB, Fleming R, Karlen M, Roberts SH (2020) Sustainable design basics. Wiley

Jeffers P (2009) Embracing sustainability—information technology and the strategic leveraging of operations in third-party logistics. Int J Oper Prod Manag 30(3):260–287

Kemp S (2015) Digital, social and mobile worldwide in 2015. http://wearesocial.net/tag/sdmw/. Accessed 12 Feb 2021

Kendall K, Kendall J (2010) Forms of government and systemic sustainability: a positive design approach to the design of information systems advances in appreciative. Inquiry 3:137–155

Kocher M (2015) Recipes and research: a survey of cookbook collection users. J Agric Food Inf 16(1):53–59

Lee S, Koubek RJ (2010) The effects of usability and web design attributes on user preference for e-commerce web sites. Comput Ind 61(4):329–341

Maceli M, Atwood M (2011) From human factors to human actors to human crafters. Paper presented at the conference, Seattle, USA

Mann S (2009) SIGCHI workshop position paper: sustainable practitioners in HCI. Paper presented at the CHI2009, Boston, USA

McDonough W, Braungart M (2002) Design for the triple top line: new tools for sustainable commerce. Corp Environ Strateg 9(3):251–258

McGovern G (2003) Usability is good management. http://www.gerrymcgovern.com/nt/2003/nt_ 2003_04_07_usability.htm. Accessed 20 Sept 2021

Miller CA, Guidry JP, Dahman B, Thomson MD (2020) A tale of two diverse qualtrics samples: information for online survey researchers. AACR

Moscardo G, Lamberton G, Wells G, Fallon W, Lawn P, Rowe A, Humphrey J, Wiesner R, Pettitt B, Don C, Renouf M, Kersham W (2013) Sustainability in Australian business: principles and practice. Wiley-Blackwell, QLD-Australia

Nathan L, Belvis E, Friedman B, Hasbrouck J, Sengers P (2008) Beyond the hype: sustainability and HCI. In: CHI 2008, Italy, pp 1–4

Newton LH (2003) Ethics and sustainability, sustainable development and the moral life. In: Basic ethics in action. Pearson Education, USA

Nicolson D, Knapp P, Gardner P, Raynor D (2011) Combining concurrent and sequential methods to examine the usability and readability of websites with information about medicines. J Mixed Methods Res 51(1):25–51

Nidumolu R, Prahalad CK, Rangaswami MR (2009) Why sustainability is now the key driver of innovation. Harv Bus Rev 87(9):57–64

Nielsen J (2003) Usability 101. http://www.useit.com/alertbox/20030825.html. Accessed 29 Sept 2021

Nies J, Pelayo S (2010) From users involvement to users' needs understanding: a case study. Int J Med Inf 79:76–82

Pan Y, Xu Y, Wang X, Zhang C, Ling H, Lin J (2015) Integrating social networking support for dyadic knowledge exchange: a study in a virtual community of practice. Inf Manag 52(1):61–70. https://doi.org/10.1016/j.im.2014.10.001

Pedol M, Biffi E, Melzi S (2021) Sustainability game. In: Corporate social responsibility and environmental management

Petre M, Minocha S, Roberts D (2006) Usability beyond the website: an empirically grounded e-commerce evaluation instrument for the total customer experience. Behav Inf Technol 25(2):189–203

Petrova E, Pauwels P, Svidt K, Jensen RL (2019) In search of sustainable design patterns: combining data mining and semantic data modelling on disparate building data. In: Advances in informatics and computing in civil and construction engineering. Springer, pp 19–26

Pichlak M, Szromek AR (2021) Eco-innovation, sustainability and business model innovation by open innovation dynamics. J Open Innov Technol Market Complex 7(2):149

Portney KE (2015) Sustainability. MIT Press

Prahalad CK, Rangaswami MR (2009) Why sustainability now the key driver of innovation. Harv Bus Rev 56–64

Preece J, Rogers Y, Benyon D, Holland S, Carey T (1994) Human computer interaction. Addison-Wesley

Preece J, Rogers Y, Sharp H (2002) Interaction design: beyond human-computer interaction. Wiley

Rahm-Skågeby J, Rahm L (2021) HCI and deep time: toward deep time design thinking. In: Human–computer interaction, pp 1–14

Ramani K (2010) Sustainable design. J Mech Des 132:1–2

Reuters T (2020) Record 53.6 million tonnes of e-waste dumped globally last year, says UN report. https://www.cbc.ca/news/science/global-ewaste-monitor-2020-1.5634759. Accessed 26 Sept 2021

Rhodes JS (2000) Usability can save your company. http://webword.com/moving/savecompany. html. Accessed 29 Sept 2021

Robertson M (2021) Sustainability principles and practice. Routledge

Sadri S, Goveas C (2013) Sustainable quality of work life and job satisfaction [an Indian case study]. J Econ Dev Environ People 2(4):26–37

Scoones I (2007) Sustainability. Dev Pract 17(4–5):589–596

Sengers P, Beale R, Knouf N (2009a) Sustainable HCI meets third wave HCI: 4 themes. In: CHI 2009, Boston, USA, pp 0–3

Sengers P, Boehner K, Knouf N (2009b) Sustainable HCI meets third wave HCI: 4 themes. In: CHI 2009, Boston, USA

Sengers P, McCarthy J, Dourish P (2006) Reflective HCI: articulating an agenda for critical practice. In: CHI'06, New York, pp 1683–1686

Shackel B (2009) Usability—context, framework, definition, design and evaluation. Interact Comput 21:339–346

Sharma A, Lyer G, Mehrotra A, Krishnan R (2010) Sustainability and business-to-business marketing: a framework and implications. Ind Mark Manag 39:330–341

Sharp H, Rogers Y, Preece J (2011) Interaction design—beyond human-computer interaction. Wiley

Shaw G, Walters R, Kumar A, Sprigg A (2015) Sustainability in infrastructure asset management. In: Proceedings of the 7th world congress on engineering asset management (WCEAM 2012), 2015. Springer, pp 525–534

Shneiderman B, Plaisant C (2010) Designing the user interface: strategies for effective human-computer interaction. Addison Wesley, Boston, MA, USA

Sikdar SK, Sengupta D, Mukherjee R (2017) Sustainability measurement for technology and business systems: use of currently available tools for quantification. Measuring progress towards sustainability. Springer International Publishing, Switzerland, pp 93–127

Silberman MS, Tomlinson B (2010) Toward an ecological sensibility: tools for evaluating sustainable HCI. In: CHI 2010, Atlanta, GA, USA, pp 3469–3474

Smith P, Sharicz C (2011) The shift needed for sustainability. Learn Organ 18(1):73–86

Smyk M, Tyrowicz J, Van der Velde L (2021) A cautionary note on the reliability of the online survey data: the case of wage indicator. Sociol Methods Res 50(1):429–464

Soden R, Pathak P, Doggett O (2021) What we speculate about when we speculate about sustainable HCI. In: ACM SIGCAS conference on computing and sustainable societies (COMPASS), pp 188–198

Stapledon T, Shaw G, Kumar A, Hood D (2015) Understanding the business case for infrastructure sustainability. In: Proceedings of the 7th world congress on engineering asset management (WCEAM 2012). Springer, pp 535–543

Stelzer K (2006) Sustainability = good design. Les Ateliters de Lethique 2:1–15

Te'eni D, Carey J, Zhang P (2007) Human computer interaction: developing effective organizational information systems. Wiley, USA

Thatchenkery T, Avital M, Cooperrider D (2010) Introduction to positive design and appreciative construction: from sustainable development to sustainable value advances in appreciative. Inquiry 3:1–14

Thiele LP (2016) Sustainability. Wiley

Tiseo I (2021) Outlook on global e-waste generation 2019–2030. https://www.statista.com/statistics/1067081/generation-electronic-waste-globally-forecast/. Accessed 26 Sept 2021

Usability First (2002) Introduction to user-centered design. http://www.usabilityfirst.com/about-usability/introduction-to-user-centered-design/. Accessed 29 Sept 2021

Vora P (1998) Human factors methodology for designing web sites. In: Chris Forsythe EGJR (ed) Human factors and web development. Lawrence Erlbaum Associates, Mahwah, NJ, pp 153–172

Wals AEJ (2014) Sustainability in higher education in the context of the UN DESD: a review of learning and institutionalization processes. J Clean Prod 62:8–15

Wang L, Kwok JS, Tsang DC, Poon C-S (2015a) Mixture design and treatment methods for recycling contaminated sediment. J Hazard Mater 283:623–632

Wang L, Tsang DC, Poon C-S (2015b) Green remediation and recycling of contaminated sediment by waste-incorporated stabilization/solidification. Chemosphere 122:257–264

Weybrecht G (2010) The sustainable MBA—the manager's guide to green business. Wiley, England, UK

Wilson P, Borras J (1998) Lessons learnt from an HCI repository. Int J Ind Ergon 22(4–5):389–396. https://doi.org/10.1016/s0169-8141(97)00093-0

Xiao Y, Faff R, Gharghori P (2013) An empirical study of the world price of sustainability. J Bus Ethics 114:297–310

Youmatter (2021) Sustainability—what is it? Definition, principles and examples. https://youmatter.world/en/definition/definitions-sustainability-definition-examples-principles/. Accessed 26 Sept 2021

Young W, Tilley F (2006) Can businesses move beyond efficiency? The shift toward effectiveness and equity in the corporate sustainability debate. Bus Strateg Environ 15:402–415

Chapter 2
Usability and Human–Computer Interaction (HCI)

Abstract Usability and HCI are becoming core aspects of the system development process to improve and enhance system facilities and to satisfy users' needs and necessities. HCI will assist designers, analysts, and users to identify the system needs from text style, fonts, layout, graphics, and color, while usability will confirm if the system is efficient, effective, safe, utility, easy to learn, easy to remember, easy to use and to evaluate, practical visible, and provide job satisfaction to the users. Adopting these aspects in the system development process, including the sustainable design, will measure and accomplish users' goals and tasks by using a specific technology. Finally, designers should include these aspects in their agenda to enhance technology acceptance, performance, and satiate users' necessaries.

2.1 Introduction

Chapter two discusses the value and the meaning of human computing interaction (HCI) and its usefulness in designing a user interface or website. "Human–computer interaction (HCI) is about designing a computer system that supports people so that they can carry out their activities productively and safely" (Preece et al. 1994, p. 1). HCI plays an important role in the development of computer systems and websites as it helps to develop "interactional techniques and to suggest where and in what situations these technologies and techniques might be put to best use" (Booth 1989, p. 6).

   Thus, a commercial website with effective HCI is likely to be more useful and profitable. HCI is a "very important concept in the system development process as it is about understanding and creating software and other technology that people will want to use, will be able to use, and will find effective when used. And the usability concept and the methods and tools to encourage it, achieve it, and measure it are now touchstones in the culture of computing" (Carroll 2002, p. xxvii). In addition, this chapter addresses the topic of usability evaluation, as usability "is concerned with both obtaining user requirements in the early stages of design, and with evaluating systems that have been built" (Booth 1989, p. 103).

© Springer-Verlag London Ltd., part of Springer Nature 2022
T. Issa and P. Isaias, *Sustainable Design*,
https://doi.org/10.1007/978-1-4471-7513-1_2

There are various methodologies to create effective websites; these methodologies address detailed issues such as page design, typography, graphics, sound, navigation, and multimedia. However, they do not provide an adequate overall approach to HCI and usability.

2.2 User-Centered System Design

In order for computer-based systems to be widely accepted and used effectively, they need to be well designed via a "user-centered" approach. This is not to say that all systems have to be designed to accommodate everyone, but that computer-based systems should be designed for the needs and capabilities of the people for whom they are intended. In the end, users should not even have to think about the complexity of how to use a computer. For that reason, computers and related devices have to be designed with an understanding that people with specific tasks in mind will want to use them in a way that is seamless with respect to their work. Additionally, it is very important to "define style, norms, roles, and even mores of human and computer relationship that each side can live with, as computers become more complex, smarter and more capable," and as we allow them to "take on autonomous or semi-autonomous control of more critical aspects of our lives and society" (Miller 2004, p. 34).

Systems designers need to know how to think in terms of future users' tasks and how to translate that knowledge into an executable system. This can be accomplished by establishing a good interface design to let the user interact and deal with the computer without any difficulties and to have more control of the system. Head (1999, p. 6) stated that good interface design "is a reliable and effective intermediary, sending us the right cues so that tasks get done—regardless of how trivial, incidental, or artful the design might seem to be."

Recently, as we know, user-centered design has become an important "concept in the design of interactive system[s]. It is primarily concerned with the design of socio-technical systems that take into account not only their users, but also the use of technologies in users' everyday activities, it can be thought of as the design of spaces for human communications and interaction" (DePaula 2003, p. 219).

HCI "is recognized as an interdisciplinary subject" (Dix et al. 2004, p. 4). HCI needs input from a range of disciplines, for example, "computer science (application design and engineering of human interfaces), psychology (the application of theories of cognitive processes and the empirical analysis of user behavior), sociology and anthropology (interactions between technology, work, and organization), and industrial design (interactive products)." Therefore, HCI has "science, engineering, and design aspects" (Hewett et al. 1992).

2.3 Human–Computer Interaction (HCI)

Before detailed consideration of the topic of human–computer interaction, two terms should be defined which are related to the development process: "Interface" and "Interaction"? According to Head, interface is the "visible piece of a system that a user sees or hears or touches" (Head 1999, p. 4). Interaction is a more general term covering the users' activity. For instance, when the user types something by using the keyboard or clicks with a mouse, this activity is called interaction.

The general concepts of HCI apply to website design. Website designers have noticed that creating a "user-friendly" site is important to maximize user response. However, designers "did[not] know any effective ways to discover what made a product user-friendly or how to design a product that was friendly" (McCracken and Wolfe 2004 p. 3). Designers often have a poor understating of HCI issues. Therefore, designers need to know how to think in terms of future users' needs, values, and supportable tasks and how to translate that knowledge into an executable system. This can be accomplished by establishing a good interface design to let the user interact and deal with the websites without any difficulties and to let the user have more control of the site.

Furthermore, in order to work effectively in the development process, HCI needs to be part of this process. According to Head, HCI has two critical dimensions in the development process: firstly, involving the user during the building and implementation of the new systems; secondly, evaluation studies about "cognitive and other behavioral factors that come into play when people interact with computers" (Head 1999, p. 9). These dimensions are consistent and mutually dependent, and thus, "the evaluation side of HCI becomes(s) a basis for decision making about design trade-offs during product development" (Head 1999, p. 9).

In the past, HCI experts tended to be consulted later in the design process, but most of the research found that this was a mistake. "The interface is not something that can be plugged in at the last minute; its design should be developed integrally with the rest of the system. It should not just present a "pretty face" but should support the tasks that people actually want to do and forgive the careless mistakes" (Dix et al. 2004, p. 3). Thus, it is important to consider how HCI will fit into the overall design process for websites (see Fig. 2.1).

2.3.1 What is HCI?

The term human–computer interaction (HCI) was adopted in the mid-1980s as a means of describing this new field of study. "This term acknowledged that the focus of interest was broader than just the design of the interface and was concerned with all those aspects that relate to the interaction between users and computers" (Preece et al. 1994, p. 7).

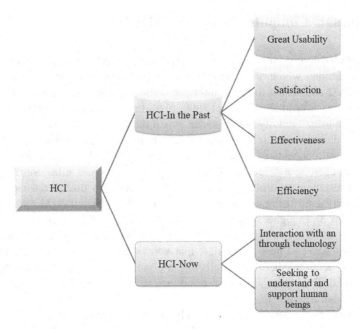

Fig. 2.1 HCI—past and now—prepared by Tomayess Issa

HCI "is a discipline concerned with the design, evaluation, and implementation of interactive computing systems for human use and with the study of major phenomena surrounding them" (Preece et al. 1994, p. 7). Therefore, the reasons for studying HCI in the development process are to create interactive computer systems that are usable and practical as well (Head 1999).

The term HCI relates to several stages in the development process, including the design, implementation, and evaluation of interactive systems, in the "context of the user's task and work" (Dix et al. 2004, p. 4). The implementation of HCI can be perceived as an art as well as a science because it requires a comprehensive range of skills, including an understanding of the user, an appreciation of software engineering capabilities and application of appropriate graphical interfaces. "If we are to be recognized as developers with professional capabilities, as competent practitioners, then it is critical to understand what makes an application interactive, instructional and effective" (Sims 1997).

HCI "is concerned with the design of computer systems that are safe, efficient, easy, and enjoyable to use as well as functional" (Preece et al. 1993, p. 11). Vora (1998) describes a framework, which provides for effective HCI for websites, with the main task being to have a clear understanding of user needs: who the users are, and what their tasks and environments are. Additionally, HCI is "concerned not only with how present input and output technologies affect interaction, but also with the consequences of new techniques such as speech recognition and generation (input and output)" (Booth 1989, p. 5).

2.3.2 HCI as Process

HCI is a discipline focusing on design, evaluation, and implementation of interactive computer systems. By adopting HCI principles and practices in the development process, the system should be easy to use by people within their work settings. The purpose of integrating HCI techniques in the overall development process is that it incorporates good design "both in practice and in understanding", and to achieve this goal, HCI addresses "what occurs on the human side of interaction as well as what happens on the machine side" (Head 1999, p. 12).

Basically, HCI is concerned with two issues: studying the relationship and the communication between the human and the computer and discovering the methods for "mapping computing functions to human capabilities and effectively using input and output techniques so that computers and users have more seamless interactions" (Head 1999, p. 12). HCI places a special emphasis on "creating and applying user-centered design techniques as well as using iterative usability testing methods" (Head 1999, p. 13).

Consequently, the machine [Computer] side involves several relevant issues including "computer graphics, operating systems, programming languages, and development environments." While on the human side, "communication theory, graphic and industrial design disciplines, linguistics, social science, cognitive psychology, and human performance are relevant" (Hewett et al. 1992).

2.3.3 Relationship Between the HCI and Human Dialogue

HCI is the study and theory of the interaction between humans and complex technology and is concerned with how current input and output technologies affect interaction, and the situations in which these technologies and techniques might be put to best use. Therefore, the relationship between HCI and human dialogue may be summarized as follows: (Booth 1989, pp. 54–55).

- Human–computer interaction, like human dialogue, is a form of communication where a degree of understanding can be achieved. Admittedly, this understanding may be limited in some respects, but if designed properly, a computer system will do as its user wishes, provided the user knows what is possible and how to give commands.
- Communications require agreement on the terms used in the dialogue. When humans successfully communicate, they usually have a shared understanding of the words used and the concepts to which they refer. This is also true of human–computer communication. When a user gives commands to a system, then the system must have an understanding of these commands if the interaction is going to succeed.
- Communications require agreement, not only upon the terms and concepts used, but also upon the context of the communication.

For example, if two people are speaking to one another, then there needs to be an agreed understanding of what they are speaking about. To illustrate this point further, let us consider an example where two individuals do not agree on the context of their conversation. Two people are sharing a car to travel to a conference. They stop at a garage for fuel and to check the car tyres. Bill is putting air into the tyres when Fred asks, "How's the pressure?" Bill replies, "Not too good, the boss keeps getting on to me." Fred explains, "Sorry I meant the car tyre pressure, but how's work anyway?" (Booth 1989, p. 55). In this example, we understand that Fred and Bill do not share a common context for their brief exchange. "In their separate contexts, the necessary link of work and the context of car maintenance, some of the words can have different meanings (i.e., "Pressure"), and the result is a failure in the dialogue between the two individuals" (Booth 1989, p. 55).

This sort of dialogue failure can also occur in human–computer communications. For example, "consider a user of a word processing system who issues a command to print the document that is currently being edited." Following the printing process, "the user issues a command for the system to re-display the document on the screen, but instead nothing happen." The system "upon receiving the first command changed to the printing mode but did not adequately inform the user who was unaware of the change in context and the subsequent legality of some of the commands." The lesson to be learned is "that those involved in communication assign [meaning] to symbols and terms depend[ing] upon the context in which they are communicated" (Booth 1989, p. 55).

The previous two examples reveal that perspective is not only important in conversation between humans, but is also a considerable factor in human–computer dialogue. To sum up, HCI is similar to human dialogue, as it is a form of communication where a degree of understanding is achieved. There must also be agreement between individuals involved in the process of communication on the meaning of the symbols and terms used. The context of the dialogue is also important, as it is the context that dictates the meanings of some of the symbols and terms used.

2.3.4 Goals of HCI

The goals of HCI are to produce usable and safe systems, as well as functional systems. These goals can be summarized as safety, utility, effectiveness, efficiency, and appeal. These goals focus on the services that the system provides, how quickly the tasks can be achieved, and ensuring that users like the system. In general, usability is an essential concept in HCI and is concerned with making systems easy to learn, easy to use, and with limiting error frequency and severity. To establish a simple system with good usability, the HCI specialists need to be aware of the following issues (Preece et al. 1994, p. 15):

- Understand the factors such as organizational, social, and psychological factors that determine how people operate and make use of computer technology effectively.
- Develop tools and techniques to help designers ensure that computer systems are suitable for the activities for which people will use them.
- Achieve efficient, effective, and safe interaction in terms of both individual human–computer interaction and group interaction.

These needs should be considered very carefully at the design stage, as most of the users should not have to change radically to "fit in" with the system; rather, the system should be designed to match their requirements.

2.3.5 Purpose of HCI

The purpose of HCI is to design a computer system to match the needs and requirements of the users. The HCI specialists need to think about the above factors in order to produce an outstanding system. To achieve the goals of HCI, a number of approaches can be utilized. These approaches need to be studied very carefully in order to develop a system, which provides the user with productivity and efficiency. These approaches are: (Preece et al. 1994, pp. 46–47).

- Involving the user: (involve the user as much as possible so that s/he can influence the system design).
- Integrating different kinds of knowledge and expertise: (integrate knowledge and expertise from the different disciplines that contribute to HCI design).
- Making the design process iterative: (testing can be done to check that the design does indeed meet users' requirements).

From the above, it was learned that HCI design should be user-centered, integrate knowledge from different disciplines, and be highly iterative. In addition, it is important to undertake effective usability evaluation. This will provide feedback regarding negative and positive aspects of prototypes.

It is important that the way in which people interact with computers is intuitive and clear. However, designing appropriate HCI is not always straightforward, as the many poorly designed computer systems testify. One of the challenges of HCI design is to keep abreast of technological developments and to ensure that these are harnessed for maximum human benefit.

The goal of this research is to develop a framework for rapid, integrated, incremental systems development that enables a group of designers and users working together to produce a friendly, effective, and efficient website. Two terms—Interaction and Interactivity—need to be defined in order to understand how the user can communicate with the system to accomplish his/her goals.

2.3.6 Interaction and Interactivity

According to Dix, "Interaction involves at least two participants: the user and the system. Both are complex, as we have seen and are very different from each other in the way that they communicate and view the domain and the task. The interface must, therefore, effectively translate between them to allow the interaction to be successful" (Dix et al. 1998, p. 104).

Users can interact with computer systems in a variety of ways. At the lowest level is batch input, in which the user provides all the information to the computer at once and leaves the machine to perform the task. This approach is called indirect interaction. An approach which involves a real-time interaction between the users and the computer is called direct interaction, as a dialogue between the user and computer will be established and at the same time will provide feedback and control right through to achieving the task.

The study of interaction can help both the HCI specialists and the users simultaneously; for example, analysis of interaction will help HCI specialists to understand exactly what is going on in the interaction and identify the likely root of difficulties. It can compare different interaction styles and take into account the interaction problems. On the other hand, the users are able to achieve their goals successfully. These goals relate to the particular application domain, i.e., an "area of expertise and knowledge in some real-world activity" (Dix et al. 1998, p. 104). The user interacts with the system for a specific reason, i.e., to perform a task, in turn to achieve the goal, which was (for instance) the reason behind visiting a particular website. So, the goal is "the desired output from a performed task" while the task is an "operation to manipulate the concepts of a domain" (Dix et al. 1998, p. 104).

To understand the interaction concept, Norman's model of interaction can be utilized (see Fig. 2.2) (Norman 1986). This model may be considered as a cycle between execution and evaluation, and these two stages can be subdivided into seven steps. The user begins the interactive cycle by defining the goal and the tasks in order to achieve his/her objectives. The user will define his/her goal by using the input mechanisms, so the task must be "articulated within the input language" (Dix et al. 1998, p. 107). Then, the input language will be translated into the system language (known by Norman as Core Language). Later, the system then "transforms itself as described by the operation translated from the input; therefore, the execution phase is complete" (Dix et al. 1998, p. 107). If the system responds to the user task in an appropriate manner to achieve the goal, then the interaction has been successful between the user and the system; otherwise, the user must "formulate a new goal and repeat the cycle" (Dix et al. 1998, p. 106).

Next, the evaluation phase begins, as the system will be in the new state and must communicate to the user the current values of the system since "attributes are rendered as concepts or features of the output" (Dix et al. 1998, p. 107). Thus, the user can see the consequences of the task s/he initiated.

Finally, is up to the user to interpret the output and to match the results of the "interaction relative to the original goal" (Dix et al. 1998, p. 107). At this stage,

Fig. 2.2 Norman's interaction model (Adopted from Norman 1986)—prepared by Tomayess Issa

the evaluation phase has ended as has the interactive cycle. A new cycle may then commence.

Norman's model is very useful as a means to understand the principles behind the interaction framework. This model allows the user to define his/her goals firstly and then will let them interact with the system to accomplish these goals. However, other researchers suggest that Norman's model considers only the "system as far as the interface and is only focusing on the user's view of the interaction" (Dix et al. 1998, p. 106). A more complex approach is needed.

The second way in which to discuss the users' communication with the system is interactivity. Interactivity can be defined in general terms as "the facility for individuals and organizations to communicate directly with one another regardless of distance or time" (Ghose and Dou 1998, p. 30). For instance, in an educational context, interactivity "refers to the activity between two organisms—which are learner and the computer" (Jonassen 1998, p. 97). In the context of HCI, "interactivity is the defining feature of an interactive system. This can be seen in many areas of HCI such as recognition rate for speech, recognition, and "feel" of a WIMP environment element: windows, icons, menus, pointers, dialog boxes, and buttons" (Dix et al. 1998, p. 136). This process is iterative with a sequence of steps and procedures followed by the user to interact with the machine (or system) to further his/her goal.

2.3.7 Factors in HCI Design

To achieve a safe and user-friendly system, the HCI specialists need to consider the main issues and factors involved in interaction and interactivity and hence in HCI design (see Fig. 2.3). These factors can be divided into (Preece et al. 1994, p. 31):

Many factors are involved, therefore, during the development process; disagreement can arise between ways to address each of these factors depending on various aspects of the system development context, such as product, team members, users, and company. According to Head (1999, p. 33), "making careful trade-offs between these numerous factors, while supporting design principles and approaches, remains a challenge of the HCI field." Consequently, most designers support involvement of the user in the design process from the beginning to reduce conflicts during the development stage.

Finally, Issa (2008) indicates that HCI is essential in the system development system. HCI will allocate users, analysts, and designers (internal and external) to identify that the website design is practical. Many specific issues need to be taken into consideration when designing website pages, such as text style, fonts; layout, graphics, and color (see Fig. 2.4).

Organizational factors	(training, job design, politics, roles, work organization)
Environmental factors	(noise, heating, lighting, ventilation)
Health and Safety factors	(stress, headaches, musculo-skeletal disorders)
The User	(motivation, enjoyment, satisfaction, personality, experience level)
Comfort Factors	(input devices, output displays, dialogue structures, use of color, icons, commands, graphics, natural language, 3-D, user support materials, mult media)
User Interface	(input device, output displays, dialogue structures, icons, 3-D, multi-medi
Task Factors	(easy, complex, novel, task allocation, repetitive, monitoring, skills, components)
Constraints	(costs, timescales, budgets, staff, equipment, building structure)
System Functionality	(hardware, software, application)
Productivity Factors	(increase output, increase quality, decrease costs, decrease errors, decrea labour requirements, decrease production time, increase creative and innovative ideas learning to new products)

Fig. 2.3 Factors in HCI—prepared by Tomayess Issa

Fig. 2.4 HCI step in the
New Participative
Methodology for Marketing
Websites (NPMMW)—Issa
2008—prepared by
Tomayess Issa

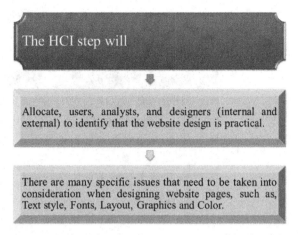

Fig. 2.4 HCI step in the New Participative Methodology for Marketing Websites (NPMMW)—Issa 2008—prepared by Tomayess Issa

2.4 What is Usability?

Usability refers to the "quality of the interaction in terms of parameters such as time taken to perform tasks, number of errors made, and the time to become a competent user" (Benyon et al. 2005, p. 52). Alternatively, usability "is a **quality attribute** that assesses how easy user interfaces are to use. The word "usability" also refers to methods for improving ease of use during the design process" (Nielsen 2003). The usability evaluation stage is an effective method by which a software development team can establish the positive and negative aspects of its prototype releases and make the required changes before the system is delivered to the target users. Usability evaluation is about observing users to "see what can be improved, what new products can be developed" (McGovern 2003). It is "based on human psychology and user research" (Rhodes 2000), since HCI specialists examine and discuss with users as they attempt and test a real task on a site (or system), and this allows them to produce a comprehensive picture of the site as practiced by the user.

From the user's perspective, usability is considered a very important aspect in the development process as it can mean the difference between performing and completing a task in a successful way without any frustration. Alternatively, if usability is not highlighted in website design, then users will become very frustrated working with it (see Fig. 2.5). For example, according to Nielsen (2003), people will leave the website: (a) if is difficult to use; (b) if the users get lost on a website; (c) the information is hard to read; d) it does not answer users' key questions; e) and lastly, if the homepage fails to define the purpose and the goals of the website. "Usability rules the web. Simply stated, if the customer cannot find a product, then s/he will not buy it. In addition, the web is the ultimate customer-empowering environment. S/he who clicks the mouse gets to decide everything. It is so easy to go elsewhere; all the competitors in the world are but a mouse-click away" (Nielsen and Mack 1994, p. 9).

Usability is a critical issue for websites as it improves competitive position, improves customer loyalty, and drives down costs (Rhodes 2000). Therefore,

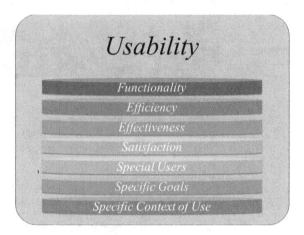

Fig. 2.5 Usability—prepared by Tomayess Issa

if usability is highlighted in website design, it will keep the organization in a powerful position compared with their competitors, as "Usability = simplicity = user satisfaction = increased profits" (Rhodes 2000).

2.4.1 Concepts of Usability

To understand fully the concepts behind the term "usability," we need to realize that usability is not "determined by just one or two constituents, but is influenced by a number of factors" which interact with "one another in sometimes complex ways" (Booth 1989, p. 106). Eason (1984) has suggested a sequence of models (see Fig. 2.6) that clarify what these variables might be. Figure 2.6 displays the relationship between independent (task, user, and system characteristics) and dependent variables (user reaction) with each variable having specific requirements and needs.

First, task characteristics are divided into frequency and openness. The frequency term refers to "the number of times any particular task is performed by a user" (Booth 1989, p. 107). If users perform a task infrequently, then help and assistance should be available via the interface so that users know which step must be taken next to accomplish the task. On the other hand, if users perform a task frequently, then it will be easier for him/her to remember the steps, which are required in order to accomplish the task.

The openness term refers to the "extent to which a task is modifiable" (Booth 1989, p. 107). This means that the information needs of the user are variable and the task must "be structured to allow the user to acquire a wide range of information." According to Eason (cited in Booth 1989), the user information needs should be fixed. If this is the situation at that time "the task need not be open and flexible, as the same information is required each time the task is performed" (Booth 1989, p. 107).

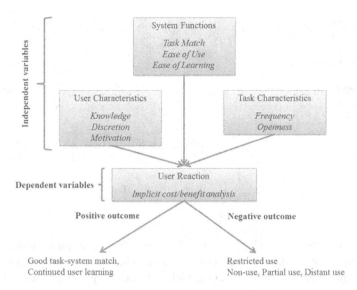

Fig. 2.6 Eason's causal framework of usability (Adopted from Eason 1984)—prepared by Tomayess Issa

The system function is described as being the most important concept under the causal framework for usability. The main concept of this variable is to improve the usability under the development process. To achieve this, the system function must address the three major system variables carefully within the development process. These are ease of learning, ease of use, and task match. The ease of learning term refers to the effort "required to understand and operate an unfamiliar system"; this term depends on the user's knowledge. The ease of use term refers to the effort that is "required to operate a system once it has been understood and mastered by the user" (Booth 1989, p. 107). The task match refers to the "extent to which the information and functions that a system provides matches the needs of the user" (Booth 1989, p. 107); in other words, whether the system will provide the necessary functions that are essential as well as the information that the user needs to accomplish his/her goals.

The final set of independent variables concerns user characteristics, focusing on who is using the system, i.e., knowledge, motivation, and discretion. Knowledge refers to the user's level of knowledge about computers and the tasks required. The motivation and discretion factors are very important concepts in the user characteristics variable with respect to the user's desire to use the system. If the user "has a high degree of motivation, then more effort will be expended in overcoming problems and misunderstandings" (Booth 1989, p. 108). On the other hand, discretion refers to the "user's ability to choose not to use some part, or even the whole of a system" (Booth 1989, p. 108). In other words, high discretion means that there needs to be satisfaction and fulfillment, via working with the new system, or the user will not bother.

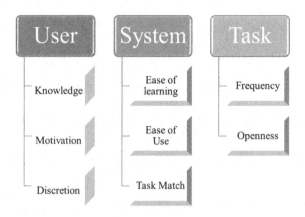

Fig. 2.7 Re-iteration of Eason's (1984) interacting task, system, and user variables—Adopted from Booth 1989, p. 109—prepared by Tomayess Issa

According to Eason (see Fig. 2.6), usability not only focuses on the user charac-teristics, but the most important aspects that need to be added in the usability chart relate to "task" and "system". Therefore, variables of task, system, and user all work jointly to establish the usability aspect of the system.

The dependent variable in Fig. 2.7 refers to the user's reaction, which Eason describes as being created by a type of cost–benefit analysis. Therefore, this variable focuses on the negative and positive outcomes of adopting the new system. Positive outcomes will lead to success of the system, while the negative outcomes will lead to suspension and discontinuation of the system. In other words, the user "accumulates a knowledge base of task-system connections as the system is used in a sequence of task episodes. The emerging strategy for use may represent a positive outcome in which the user locates and uses appropriate system functions for every new task and progressively masters the system. The reverse scenario occurs when negative outcomes prevail and use of the system is discontinued. Eason points out, based on his field studies, that under realistic conditions the user appears to approach a state of equilibrium where further learning about the system is minimized" (Lowgren 1995, p. 5).

2.4.2   Usability Criteria

Various principles need to be followed in order to support usability, making systems easy to learn and easy to use. These principles are (Dix et al. 1998, p. 162 and Nielsen 2003):

- *Learnability*: by which new users can begin effective interaction and achieve maximal performance.
- *Flexibility*: the multiplicity of ways the user and system exchange information.
- *Robustness*: the level of support provided to the user in determining successful achievement and assessment of goals.

- *Efficiency*: once the user learns about the system, [the speed with which s/he] can perform the tasks.
- *Memorability*: how easily the user will remember the system functions, after a period time of not using it.
- *Errors*: "How many errors do users make, how severe are these errors, and how easily can they recover from the errors?" (Nielsen 2003).
- *Satisfaction*: how enjoyable and pleasant is it to work with the system?

These principles can be applied to the design of an interactive system in order to promote its usability. Therefore, the purposes behind adopting these principles are to give more assistance and knowledge to system developers (and the users) regarding the system design. Alongside the above principles, an important key additional factor is utility. Utility refers to the functionality so users can "do what they need or want to do" (Preece et al. 2002, p. 16). In other words, "does it do what users need?" (Nielsen 2003). For that reason, usability and utility are equally important in the development process, and they need to be integrated.

2.4.3 Usability Specifications

Once the designer has gathered and analyzed information about the tasks, problems, and steps to work with the proposed system, the next step is to answer the question: How will we know if the interface is usable? This is laid out in a usability specification.

A usability specification defines the measure of success of a computer system or website and serves as an indicator about whether or not the development of the website is on the right track. A usability specification should be developed during the first stage of the development process and monitored "at each iteration", to determine whether the "interface, is, indeed, converging toward an improved, more usable design" (Hix and Hartson 1993, p. 222). Usability specifications should lay out explicitly how usability will be evaluated and can be divided into two sections:

- *Performance Measures*: are directly observable by watching a user complete a task within a specific time. This includes monitoring the number of errors and time needed to accomplish the task. These types are "quantifiable measures" which means that they can be communicated with numbers. For example "you can count the number of minutes it tasks a user to complete a task or the number of negative comments that occur" (McCracken and Wolfe 2004, p. 53).
- *Preference Measures*: give an indication of a "user's opinion about the interface which is not directly observable" (McCracken and Wolfe 2004, p. 53). Preference measures can be determined by using questionnaires or interviews.

Usability specifications are needed to determine when the iteration of prototypes has produced a system with sufficient usability. Therefore, without usability specifications, the key factors that "generally determine an end to the iterative refinement process are when developers run out of time, patience, and/or money" (Hix and

Fig. 2.8 Usability step in the New Participative Methodology for Marketing Websites (NPMMW)—Issa 2008—prepared by Tomayess Issa

Hartson 1993, p. 243). Usability specifications are very important to the development process since they define "a quantifiable end to the seemingly endless iterative refinement process" (Hix and Hartson 1993, p. 242).

Lastly, Issa (2008) confirms that usability is a core step in the system development process as usability will allow users, analysts, and designers (internal and external) to confirm that the website design is efficient, effective, safe, utility, easy to learn, easy to remember, easy to use and to evaluate, practical, visible, and provide job satisfaction (see Fig. 2.8).

2.5   Conclusion

This chapter has outlined the basic concepts involved in human–computer interaction and usability in the system development process. These considerations are very useful to the business community in line to increase the efficiency of their staff and thus their profits. Currently, HCI and usability are needed in any design, including sustainable design to recognize the new smart technology and portable device needs from designers and users' perspective. Therefore, designers should integrate and combine HCI and usability in their agenda design, including sustainable design, to enhance new smart technology and portable devices performance and facilities, and to satisfy users' needs.

References

Benyon D, Turner P, Turner S (2005) Designing interactive systems: a comprehensive guide to HCI and interaction design, 2nd edn. Pearson Education Limited, UK

Booth P (1989) An introduction to human-computer interaction. Lawrence Erlbaum Associates, Publishers, Hove, East Sussex, UK

Carroll JM (2002) Human-computer interaction in the new millennium. Addison-Wesley, New York

DePaula R (2003) A new era in human computer interaction: the challenges of technology as a social proxy. In: Latin American conference on HCI, Rio de Janeiro, Brazil 2003. ACM international conference proceeding series, pp 219–222

Dix A, Finlay J, Abowd G, Beale R (1998) Human-computer interaction, 2nd edn. Pearson Education Limited, USA

Dix A, Finlay J, Abowd G, Beale R (2004) Human-computer interaction, 3rd edn. Pearson Education Limited, England

Eason KD (1984) Towards the experimental study of usability. Behav Inf Technol 3(2):133–143

Ghose S, Dou W (1998) Interactive functions and their impacts on the appeal of Internet presence sites. J Advert Res 38(2):29

Head AJ (1999) Design wise. Thomas H Hogan Sr.

Hewett T, Baecker R, Card C, Carey T, Gasen J, Mantei M, Perlman G, Strong G, Verplank W (1992) Human-computer interaction. ACM SIGCHI curricula for human-computer interaction. http://old.sigchi.org/cdg/. Accessed 5 Oct 2021

Hix D, Hartson HR (1993) Developing user interfaces: ensuring usability through product and process. Wiley, USA

Issa T (2008) Development and evaluation of a methodology for developing websites. PhD Thesis, Curtin University, Western Australia. http://espace.library.curtin.edu.au/R/MTS5B8S4X3B7SBAD5RHCGECEH2FLI5DB94FCFCEALV7UT55BFM-00465?func=results-jump-full&set_entry=000060&set_number=002569&base=GEN01-ERA02.

Jonassen DH (1998) Instructional designs for microcomputer courseware. L. Erlbaum Associates, Hillsdale, NJ

Lowgren J (1995) Perspectives on usability. Department of Computers and Information Science, Linkoping University, Sweden. http://www.ida.liu.se/labs/aslab/groups/um/publications/R-95-23.PDF. Accessed 5 Oct 2021

McCracken DD, Wolfe RJ (2004) User-centered website development: a human-computer interaction approach. Pearson Education Inc., New Jersey, USA

McGovern G (2003) Usability is good management. http://www.gerrymcgovern.com/nt/2003/nt_2003_04_07_usability.htm. Accessed 20 Sept 2021

Miller CA (2004) Human-computer etiquette: managing expectations with intentional agents. Commun ACM 47(4):31–34

Nielsen J (2003) Usability 101. http://www.useit.com/alertbox/20030825.html. Accessed 29 Sept 2021

Nielsen J, Mack RL (1994) Usability inspection methods. Wiley, NY

Norman DA (1986) Seven-stage model of (individual) interaction. Department of Computer and Systems Sciences, Lulea University, Bai Guohua, Sweden

Preece J, Rogers Y, Benyon D, Holland S, Carey T (1994) Human computer interaction. Addison-Wesley, England

Preece J, Rogers Y, Keller L, Davies G, Benyon D (1993) A guide to usability "human factors in computing". In: Preece J (ed). Addison Wesley, England

Preece J, Rogers Y, Sharp H (2002) Interaction design: beyond human-computer interaction. Wiley, USA

Rhodes JS (2000) Usability can save your company. http://webword.com/moving/savecompany.html. Accessed 29 Sept 2021

Sims R (1997) Interactivity: a forgotten art? http://www.gsu.edu/~wwwitr/docs/interact/. Accessed 5 Oct 2021

Vora P (1998) Human factors methodology for designing web sites. In: Chris Forsythe EGJR (ed) Human factors and web development. Lawrence Erlbaum Associates, Mahwah, NJ, pp 153–172

Chapter 3
User Participation in the System Development Process

Abstract User participation in the system development process is crucial and vital to ensure if user interfaces, devices including website are successful and easy to learn and implement as user participation will improve and enhance performance and increase user acceptance and satisfaction. User participation will encourage users to participate in decision making and actions during the system development process. The user participation rational will reduce the time taken by designers in various stages from implementation, testing, evaluation, and training, since users will become more aware behind the new design. This chapter aims to discuss the importance of user participation in the system development and sharing with the readers the why, how, and when we need to involve participants in the design process.

3.1 Introduction

This chapter focuses on users, their work, and their environment and the reasons for involving them in the design process. Participation role in the system development process is crucial and critical to ensure if the design process will be successful or unsuccessful. In general, if designers manage to work very closely with the users to produce new smart technology or portable devices, then less time will be required in the implementation, testing and training stages, and consequently, the user will work with the new devices, with less frustration and dissatisfaction. This chapter is organized as follows: What is Participation, How We Know Our Users, and Conclusion.

3.2 What is Participation?

Participation is "A process in which two or more parties influence each other in making plans, policies or decisions, it is restricted to decisions that have future effects on all those making the decisions or on those represented by them" (Mumford 1995, p. 12). It can also be defined (in the context of systems development practices) as the

© Springer-Verlag London Ltd., part of Springer Nature 2022   41
T. Issa and P. Isaias, *Sustainable Design*,
https://doi.org/10.1007/978-1-4471-7513-1_3

"extent to which the user engages in systems analysis activities such as project defini-tion and logical design decisions" (Doll and Torkzadeh 1989, p. 1154). Furthermore, user participation is defined as the "behaviors, assignments, and activities that users or their representatives perform during the information system development" (Hartwick and Barki 1994, p. 441). A high level of user participation is likely to enhance user "ownership" of, or identification with the resulting system—in this sense "'user involvement' refers to the set of all such user subjective attitudes toward, or psycho-logical identifications with, information systems and their development"(Kappelman 1995, p. 70). However, the term "user involvement" can also refer to a low level of participation, where users have little power to influence decisions.

This research focuses on "user participation" not "user involvement" as the former term implies a role for the users which is more powerful and influential in the devel-opment process, especially in website design, as the user will be actively engaged throughout the development process. This will assist the user to accept and compre-hend the system. Participation is more "effective when an individual's desire or "motivation to participate" is in congruence with perceptions of actual involvement" (Doll and Torkzadeh 1991, p. 443). Decisions about the role of the user need to take into consideration that users are "becoming more knowledgeable and active in defining their information requirements" (Doll and Torkzadeh 1989, p. 1154).

This research distinguishes between two types of users: end-users (internal to the client organization) and client–customer users (external). End-users (Internal) are the real users in the client organization, who test and evaluate the website and use it to respond to the client–customer's queries. The client–customer users (external) are those who interact with the website to accomplish their goals such as purchasing goods or services from the client organization. It is important to understand the needs, desires, and characteristics of both types of users. To date, most designers of websites have "assumed that their users had the same background and expectations that they did"; therefore, "the more you know about your users and their work, the more likely it is that you will develop a usable and successful website" (McCracken and Wolf 2004, p. 37). These two types of users (see Fig. 3.1) should both participate in the development process under the methodology developed during this research to make sure that the website meets the requirements of end-users, client–customers, and designers simultaneously. The purpose behind this participation has various benefits: (1) to reduce the time in the implementation and testing stages; (2) to familiarize the end-users and client–customers with the new system before the implementation; (3) and provide job satisfaction and meet the task effectiveness needs of the end-users and client–customers.

User participation assists system development by providing a "more accurate and complete assessment of user information requirements, providing expertise about the organization the system is to support, expertise usually unavailable within the information systems group, avoiding development of unacceptable or unimportant features and importing user understating of the system" (McKeen et al. 1994, p. 427–428). Tait and Vessey stated that participation "reduces the risk of system failure in complex projects" (cited in (Amoako-Gyampah and White 1993, p. 2)). Therefore, in order to make the system more successful, participation needs to be an integral

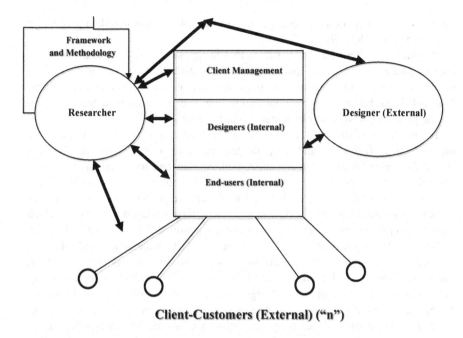

Client-Customers (External) ("n")

Fig. 3.1 Users (end-user and client–customers)—Prepared by Tomayess Issa

part of "the design and implementation process" (Tait and Vessey 1988, p. 91), not just a convenient add-on.

Participation in the development process can be "viewed as "sharing" in decision making or engaging in activities" (Doll and Torkzadeh 1989, p. 1155) and to determine "information requirements by encouraging users and other to indicate what they do and what information they need to do it" (Hepworth et al. 1992, p. 122). Research has shown that user participation in system design will greatly assist in producing a successful system. It results in less time in the implementation and testing stages as users are more knowledgeable about the system.

The user's participation is very important since the lack of "user involvement as the chief reason IS projects fail" (Engler 1996, p. 3), and "developing an information system without user participation tends to result in the delivery of systems that fail to meet the users' needs" (Hawk and Dos Santos 1991, p. 317). After reviewing the role of user participation in different types of projects, Hirschheim (1985) asserts "more user participation was undertaken by organizations when the systems were complex" (cited in (Amoako-Gyampah and White 1993, p. 2).

User participation should be introduced in the development process to ensure that the system is successful and easy to implement as user participation may lead "to improved system quality as well as increased user acceptance, reflected in increased use of and satisfaction with the system" (Baroudi et al. 1986, p. 233). In addition, it will decrease resistance and increase acceptance of planned change (Baroudi et al.

1986). User participation will change "the attitude of user towards data processing and vice versa" (Doll 1987, p. 27).

Research and experience have shown that to run a successful application development process without any frustrations and dissatisfaction, the designer needs to involve the users and set clear objectives and recognition of organization factors. This will help the designer incorporate the views of users in all of the following stages: planning, design, implementation, and testing.

To implement such an approach, a designer may adopt, for instance, the Effective Technical and Human Implementation of Computer Systems (ETHICS)methodology, as it considers both human and technical factors when designing a new system. In other words, this is known as a "socio-technical" approach, which "recognizes the interaction of technology and people, and produces work systems which are both technically efficient, have social characteristics which lead to high job satisfaction and create high quality products" (Mumford 1995, p. 2).

Before adopting this approach, a designer needs to understand, and take into account, that each user will have different characteristics, such as interest, values, and needs. These considerations need to be met by both parties—employee and the management to "accept major change willingly and enthusiastically" (Mumford 1995, p. 2). Some researchers indicate that some organizations will let the management play a large role in developing a new system, while the users will participate in a small way, or sometimes they will not participate at all. Hence, user participation can be at various levels and in different ways. According to Tait and Vessey ((Cited in (Saleem 1996, p. 147)), there are various types of participation, for example:

- *No participation*: users are not invited to participate.
- *Symbolic participation*: user input is sought but ignored.
- *Participation by advice*: users are consulted.
- *Participation by weak control*: users may have sign-off responsibility.
- *Participation by doing*: users are members of design team:
- *Participation by strong control*: users may pay for the system development.

The use of options involving little user participation will create numerous problems for the users as well as the management, as users will most likely find that this system is not meeting their needs, desires, and is very hard to cope with. This may lead to "serious morale problems" (Mumford 1995, p. 2) resulting in reduced job satisfaction, low efficiency "low commitment to the system, together with increased resistance to any future change" (Mumford 1995, p. 2).

3.2.1 Change Processes

To be successful and meet user requirements, the development of a new system requires a number of "change process" aspects to be considered by the designer, user, and management simultaneously. These aspects are objective setting and attainment; adaptation; integration; and stabilization.

- **Objective Setting and Attainment**: this should involve all the groups (not only the senior management) from an organization who intend to use the system. Each group (or every individual) will have special interests and values. Consequently, designing a system for today and the future needs to involve various sessions of brainstorming between the users to exchange opinions and views to enhance the system. Today "non-technical users are familiar with, and knowledgeable about, the advantage and disadvantages of technical systems" (Mumford 1995, p. 6). Users are "becoming more sophisticatèd and as they do so, their expectations and behaviors are changing. Don't get caught designing for yesterday's audience— stay on the cutting edge with this kind of research so that you can design for tomorrow's audience!" (Sheridan 1999). Moreover, these groups are "able to make informed choices on the hardware and software that will best meet their needs" (Mumford 1995, p. 6).

- *Adaptation*: this process is "moving from one kind of technical and organizational structure and state to another, and the means by which this change is assisted to take place smoothly and successfully" (Mumford 1995, p. 7). Adaptation occurs in the implementation phase of the new system. The adaptation needs to address issues such as values, interests, attitudes, motivations, and the conflicts between the groups who are working together to implement a new system. Therefore, support and assistance need to be provided from the top management to understand and study any potential conflicts between groups of users. This step is very significant to reduce any struggle between the groups and to certify that the system is running smoothly, according to the users' needs.

- *Integration*: "is the action taken, once the system has been designed and is being implemented, to ensure a new situation reaches a state of equilibrium" (Mumford 1995, p. 7). The purpose behind integration is to gather different aspects such as task, technology, people, and organizational environment into a valuable relationship between themselves. The relationship between these aspects should be stable and capable of adoption. Organizations should respond directly to all the changes which occur in their environment "while at the same time either maintaining a state of equilibrium or being able to make adjustments which restore equilibrium if internal relationships are distributed" (Mumford 1995, p. 8). Introducing a new technology to the above aspects (task, technology, people, and organizational environment) will bring a new relationship between them, which should integrate "both opportunities and constraints" (Mumford 1995, p. 8). Since tasks are influenced by technology, the task structure of "functions or departments using the system will be altered" (Mumford 1995, p. 8). New tasks will have new demands; therefore, in this scenario, job satisfaction will be affected, as new tasks will have new demands and requirements that will produce negative or positive feedback. Consequently, technology, people and tasks will interact with the environment to provide a new structure "for the achievement of the organization's objectives and interaction may start the looping process again by making new demands of technology" (Mumford 1995, p. 8). Thus, integration requires adaptation in order to produce a good relationship between technology, people, tasks, and organizational structure.

- *Stabilization*: this is the last step in the change process. Stabilization requires that "once new patterns of behaviour have been successfully initiated; they must be established and reinforced" (Mumford 1995, p. 6). This means that the relationship between the aspects (task, technology, people, and organization) should incorporate the new patterns of task performance, which is required by the system to ensure that they meet the values and interests of groups who are involved.

In summary, designers need to take into account the above change processes during the development process of a new system, and these changes should be considered from the human perspective, not from the technical aspect. This means that user participation should be a priority from the beginning, involving the user in all stages of the process from planning to implementation. This action will achieve two desirable outcomes: a successful system and job satisfaction.

Previously, users were involved only in the analysis and design phases, as most of the methodologies are "designed around the needs and capabilities of analysts instead of users" (Dean et al. 1997, p. 186). Nevertheless, these days users should be involved from the beginning to the end as he/she will be able to interact with the system more and to provide more feedback to support effective iteration at each step.

Designers need to select as participants the users who are dealing with the system on a daily basis, not the management and technical personnel. The human aspect has the positive aim of "encouraging the setting and achieving of human objectives as an integral part of the design process" (Mumford 1995, p. 11).

3.2.2 Managing User Participation in Development Processes

Before adopting a participative approach to system development, it is very important to estimate the functions, structures, and processes of participation and to understand the relationship between the management, technical personnel and finally, the more important source, the users. Participation can play a significant role in promoting and endorsing the development process, as participation will "lead to successful outcomes in terms of more information system usage, greater user acceptance, and increased user satisfaction" (Lin and Shao 2000, p. 283). Indeed, "participation is morally right – people should be able to determine their own destinies" (Mumford 1995, p. 13). It enables users to learn more about the system before implementation, producing an "interested and committed group of staff and therefore assisting in the avoidance of morale and job satisfaction problems" (Mumford 1995, p. 13).

Typically, user responsibilities in the participation stage will extend from the beginning until the end of the development process, including the testing and evaluation of the system. For example, user responsibilities can involve "project initiation, determining system objectivities and information needs, identifying sources of information, analyzing information flows, developing input and output formats/screens, and specifying aspects of the user interface" (Doll and Torkzadeh 1989, p. 1155).

Participation is considered a valuable experience for some users who will be involved in the system development process since they will obtain more knowledge, experience about the system before it is implemented. Furthermore, Hartwick and Barki (1994) indicate that users who participate in the system development process are likely considered that the new system is important and good.

Users will be interested in and attracted to the participation process, as it will:

- ***Enable them to*** "prevent things that they believe to be undesirable from happening."
- ***Avoid and prevent the*** "users to undertake tasks that they regard as time-consuming and irrelevant or even being made redundant".
- ***Help the users*** to make their job more interesting, providing "better services to the client-consumers, promotion, and improved quality of working life";
- ***Enhance group harmony***, as it develops a "sense of cooperation and community and produces a willingness to accept group decisions."

(Mumford 1995, p. 13)

Although these theories of participation have been primarily developed in the context of design of information systems, they apply equally to the development of websites. Merrick (2001, p. 67) states, "it's important to reach online-users because they are generally the most profitable."

3.2.3 How to Participate?

Participation has a different significance and sense for different groups and individuals, as they have different objectives. Management and designers need to act as a team to present a set of processes and structures that will help the users to achieve their objectives. These gains "will not necessarily be all of the same kind but they should enable each group to say with conviction "participation has clear benefits for us"" (Mumford 1995, p. 13).

The participation process needs to be examined very carefully by both parties (designers and management) to decide which participative approaches should be adopted for the particular development process. There are two main types of participation: indirect "where user representatives participate in the system development process"; and direct "where the users themselves fully participate in the development process" (Barki and Hartwick 1989, p. 54).

Each participation type has special techniques and particular requirements when it is adopted for the development process. For example, if the indirect approach is chosen, then the most important issue that needs to be addressed is to ensure that all interests are represented. Users should decide "how the members of the participative forum are selected or elected and whether a number of groups at different organizational levels are required" (Mumford 1995, p. 14). While, if the direct participation approach is adopted in the development process, the designers and management need to define various issues at the beginning; for example, the degree of participation and the degree of influence that users will have regarding changing aspects of the design, before the implementation.

Users can play a significant role in the development process and this involvement and participation can be in the beginning, middle or at the end of the development process. Each step of this participation has specific requirements and procedures that must be followed so that users can play their role in developing the new system, with anticipation that it will meet their desires and requests.

Mumford (1995) provides a slightly more complex model of participation options. She notes three types of involvement: consultative, representative, and consensus. Each one has specific requirements from the users and designers' perspectives.

- *The Consultative approach*: is very useful to secure agreement and settlement between the users and designers at the beginning, to define the objectives of the new system. This approach will allow the full hierarchy of people (top, senior, and low management and interested subordinate staff) to work together to define organizational future needs with respect to the new system. However, "consultative structure must exist or be created so that this sounding out of opinion can be thorough and accurate" (Mumford 1995, p. 18).
- *The Representative approach*: is very appropriate at the definition stage. It is considered useful and powerful since a hierarchy of people will contribute to system definition and setting the boundaries of the new system. A representative approach requires input from all the functions and levels in those parts of the organizations that are using the information system. The design group "will see an important part of its task as involving its departmental colleagues in the design activities and in the decision taking on how work is to be reorganized around the technical system" (Mumford 1995, p. 18).
- *The consensus approach*: is more popular in most organizations as it enables all the staff associated with developing a new system to take part and have a role in designing the new system for an organization. This is achieved "when efficiency and job satisfaction needs are being diagnosed through feedback and discussion in small groups" (Mumford 1995, p. 18).

It is important to note that each approach has specific time constraints, needs, activities, and potential problems. For example, the consensus approach "does not always emerge easily, and conflicts which result from different interest within a department may have to be resolved first" (Mumford 1995, p. 19). Hence, the other approaches (representative or consultative) are often adopted when developing a new system for an organization.

A participative approach is very useful at all stages, as it will "lead to efficiency gains, the creation of high-quality customer care and a good work environment, and more job satisfaction for staff" (Mumford 1995, p. 19). According to Mumford, two types of groups should carry out the stages in the process of systems development (i.e., planning, design, implementation, and evaluation):

- *The first group* is responsible for steering the project. The purpose of this group is to provide the link between the different people involved in the project. Moreover, the role and responsibility of this group is to define the "objectives and constraints under which the new system is to be developed" (Mumford 1995, p. 19).

- *The second group* is responsible for defining the system design, to support the function or department where the new system will be implemented and introduced. The role and responsibility of this group is to define the problem, environment, system goals, and (the most important aspect) to identify the impacts of the new systems at each level in the organizational hierarchy.

User participation during the system design will lead the user to understand more about the system firstly, and hence, the system will be more productive and efficient. User participation will "improve the quality of design decisions and resultant applications, improve end-user skills in system utilization, develop user abilities to define their own information requirements, and enhance user commitment to and acceptance of the resultant application" (Doll and Torkzadeh 1989, p. 1152). Moreover, "user satisfaction with a system is a component of job satisfaction, one would anticipate a positive relationship between user involvement and user satisfaction" (Lawrence and Low 1993, p. 196). Participation by users in the development process will provide a more accurate and complete assessment of user "information requirements, avoiding development of unacceptable or unimportant features; improving user understanding of the system and finally will lead to decreased user resistance" (Amoako-Gyampah and White 1993, p. 2).

Rondeau et al. (2002, p. 151) stated that "involving product development managers and manufacturing managers (i.e., end-users) in IS-related activities enables firms to build an IS infrastructure that supports cross-functional decision making." System requirements information can be obtained from the user by using the interview method. This method should be introduced in the development process of websites to gain more information about the "basic content areas of the site" (Fleming 1998, p. 213). Consequently, to meet the user needs, Fleming (1998) suggests that a three-tiered system of goals-(basic), purpose-(oriented), and topic (or audience) should be considered. The basic goals relate to navigation questions such as "Where am I?" Or "Where can I go?" (Applen 2002, p. 305). Moreover, such design approaches should involve user participation. Effective "communication and positive relationships must be cultivated and planned as any other successful component of project management" (Jiang et al. 2002, p. 20). According to Engler (1996, p. 72), these are the steps, which need to be followed, by designers and management simultaneously during the development process:

- *Identify the correct user*: throughout this step, the designer will define the full range of users and plan for gaining customer input, not just internal user input.
- *Involve the user early and often*:

  - Get the user involved in the development process at all stages (i.e., development, implementation, and maintenance).
  - Rules and procedures should be established to motivate the users during the development process.

- – Educate and negotiate with the users regarding their roles and responsibilities— "listen to the users' expectations, what does "involvement" mean to them." (Engler 1996, p. 72).
- – Assign a Facilitator who comprehends the required relationship between designers, management, and the users. On other words "someone with a foot in both worlds" (Engler 1996, p. 72).

- • *Create and maintain a quality relationship*: this step can be achieved by meeting, understanding, and listening very carefully to the users.
- • *Make improvement easy*: finally, the designer needs to learn the following concepts with respect to the users:

  - – Learn the user's language.
  - – Proactively solicit the user's opinions.
  - – Show the user that his/her opinions make a difference.
  - – Make sure there's a demonstrated benefit for user involvement.

3.2.4   Some Problems with the Participative Approach

A participative approach is very practical and valuable to the designer and users simultaneously. It is considered "an important mechanism for improving system quality and ensuring successful system implementation"(Baroudi et al. 1986, p. 232) and "is used to gather local intelligence about particular needs and difficulties at different project sites" (Kawalek and Wood-Harper 2002, p. 18).

However, some system developers believe that a participative approach will create problems for the people who are involved in it, especially to the users. Participation in the system's development process can be seen as "manipulative, will impair labor shedding, will entrench poor practice, can lead to poor design, is not cost-effective, and can be dysfunctional because it can lead to political problems" (Lawrence and Low 1993, p. 195). Hirschheim (1985, p. 295) states that participation can lead "to systems which are not only sub-optimal, but take much longer to develop, and is extremely difficult to operationalize."

According to Mumford (1995), a participative approach can create a few problems for some of the people who are involved in the development process, particularly the users., for example decrease in trust, conflict over election versus selection of representatives, conflicts of interest, and stress. Key issues include communication and consultation; professional systems designer's role; and finally, the functional or departmental manager role. These problems can occur if the management did not determine the desires and requirements of the people who are involved in the development process, particularly the users.

To prevent and resolve these conflicts, the management needs to address two objectives: a) firstly, establish good communication mechanisms—for instance, establish a weekly group meeting to provide consultation and commutations skills; and b) secondly, the management must be in continuous contact with the users to confirm whether or not they are on the correct track with the development process. All problems need to "be recognized, brought out into the open, negotiated and a solution arrived at which largely meets the interest of all parties in the situation" (Mumford 1995, p. 25). Finally, Olson and Ives (1981) stated that "much of the existing research is poorly grounded in theory or methodologically flawed; as a result, the benefits of user involvement have not been convincingly demonstrated" (Cited in Hirschheim 1985, p. 295).

3.3 How We Know Our Users

This section will discuss the following aspects: defining who the users are in general; user's goals, activities, and environment; their special effects on usability specifications; and the techniques for observation of, and listening to, users.

Users include "those who manage direct users, those who receive products from the system, those who test the system, those who make the purchasing decision, and those who use competitive products" (Preece et al. 2002, p. 171). The different types of users are very important concepts in this research as, through them, the interface can be developed in a way, which meets their needs.

The rationale behind involvement of users in website development is: 1) to reduce time in implementation and testing stages; 2) to familiarize the end-users and client–customers with the new system before the implementation; and 3) provide job satisfaction and meet the task effectiveness needs of the end-users and client-customers. A user-centered, task-based approach to system development is required as both user and task analysis needs must be determined and analysed very clearly at the beginning of the development process, to prevent any problems with respect to high maintenance costs and user frustration. For example, to make the business booming and prosperous, the supplier needs to answer and meet user requirements regarding services, products, and prices.

3.3.1 User Characteristics

In order to design effectively for users, there are a few user characteristics, which need to be defined for any web project, such as "Learning style, tool preference, physical differences, and cultural differences" (McCracken and Wolfe 2004, p. 38). Unless the system is customizable by the users, then it is the "average" or, most likely, characteristics of the target user population which need to be considered.

- *Cognitive and Learning Style*: Users will have different cognitive and learning styles. For instance, it is useful to distinguish between the user types "'read then do' people or 'do then read'" people (McCracken and Wolfe 2004, p. 38). In other words, do your users want and expect full instructions before starting or do your users directly work with the interface without any help and instructions?

- *Interface/Interaction Preferences*: the developer also needs to define user differences with respect to their preferred web interaction techniques (pull-down menu, Windows, etc.) and pre-fined mode of interaction with the interface (mouse or keyboard). Other questions which need to be asked about the users include:

  - What computers, interfaces, and browsers are users currently using?
  - Do they always use the same ones or are they familiar with a range of versions?
  - Where did they learn these tools? School? On-the-job training? On their own?
  - How familiar are they with the tools? How often do they use them? When did they learn?
  - Are they familiar with technology that is similar to your intended design? Do they understand frames? Pop-up windows? Search commands?

(McCracken and Wolfe 2004, p. 39)

Besides the above information, the designer needs to learn more about the user's knowledge and background in dealing with the interface; for example, are the "users just starting to use the Internet?" (McCracken and Wolfe 2004, p. 39). If they are novices, it is better to observe them and to assess whether the interface will cause problems and frustration. This experience will help the researcher to find out about problems, which could cause frustration, and how these issues can be resolved before the implementation. Other user classifications relate to:

- *Physical Differences*: The designer needs to gather more information about the typical user, such as age, gender, color blindness, and other physical disabilities.
- *Application Domain Differences*: the designers should also collect more information about the background of their users. For example, if the designer needs to design a website for education, then the vocabulary is different from that used for users from different applications domains—dentists, architects, or bankers, and so on. According to McCracken and Wolfe (2004, p. 41) "What the 'default'" means to a banker is different from what it means to a programmer. Using the appropriate vocabulary will prevent the user from being forced to ask, "Is this the link I want?" and will empower the user with the conviction, "I want this link."

From all the possible types of user characteristics, a particular set of user classifications (taxonomy) must be selected for a specific website project. For instance, Turk (2001, p. 163) recommends consideration of the following key user characteristics:

- Age
- Culture
- Disabilities
- Education level
- WWW/IT experience

The designer should consider these various user characteristics in relations to the design of the website, i.e., the level or particular option for each characteristic— for the average user (and the range) for the target user population. Moreover, more questions need to be asked of the users with respect to visiting a website, for example: the purpose behind visiting this website, how they will work with it, and if they are familiar with this website or ones similar to it. These questions will help the designer to gain more information about the users' knowledge of websites.

3.3.2 Knowledge of User Tasks

This stage in the design process focuses on the purpose behind using the website. For example, if the website is part of a formal work procedure, the designer could expect that the users will be well trained to work with the website. The designer also needs to know if their website-based activities will fit into the workflow of the users' business, and they need to understand "what has been done before the work gets to them, and do they know what happens afterwards" (McCracken and Wolfe 2004, p. 42).

Consequently, designers should understand and recognize two things before they work with the users. Firstly, the designer needs to know the purpose behind visiting the website—is it (for instance) to gain information, shopping, or entertainment? Secondly, the designer needs to gain more information about the users' job and the degree of "familiarity they have [with] the basic tools of technology" (McCracken and Wolfe 2004, p. 42).

McCracken and Wolfe (2004) suggest that it is important to understand the users' level of expertise. Users with the lowest level of expertise are termed "Novices." This type of user is "learning a skill for the first time." Novices have a poor understanding of the parts of the website and typical use scenarios. Novices "only recognize a few positions and have not developed any such sequences" (Preece et al. 1994, p. 163). As a result, the purpose of visiting the website is often just to complete a particular task, which they believe will achieve their goals. More advanced users may be classified as follows:

- *Advanced Beginner*: this type of user "is focused simply and exclusively on getting a job done as painlessly and quickly as possible" (Hackos and Redish 1998, p. 82). These people are at the developing stage of expertise, and they have knowledge of how to deal with this application and to go through it without any tribulations, especially when the steps are direct and easy to follow. However, these users will be very confused if there are many alternatives to choose from, and if they "encounter difficulties, they have trouble diagnosing or correcting the problem" (McCracken and Wolfe 2004, p. 43).
- *Competent Performer*: these types of users are those "who have learned a sufficient number of tasks that they have formed a sound mental model of the subject matter and the product" (Hackos and Redish 1998, p. 84). These people are willing to

learn and study by themselves the principles of how to work with this website. These people may prefer working with the website (or system) via a user manual and documentation to accomplish their goals.

- *Expert*: these users "perform the task automatically without consciously having to think about each move" (Preece et al. 1994, p. 163). These people have the knowledge to perform a wider range of complex tasks and "suggest solutions to problems" (Preece et al. 2002, p. 346). Experts can develop a "repertoire of sequences of moves" (Preece et al. 1994, p. 163), unlike the novices who are able to utilize only a small set of use scenarios.

Preece et al. (2002) provide a further way of classifying users: the "Primary users" who are likely "to be frequent hands-on users of the system," while the "Secondary users" are "occasional users or those who use the system through an intermediary, and "Tertiary" users are those who are affected by the introduction of the system or who will influence its purchase" (Preece et al. 2002, p. 171).

3.3.3   Recruiting Users

With regard to users, "a representative sample must be involved throughout the design process, from the very beginning" (Cato 2001, p. 41), as they can help the designer not only in one stage but in all the stages. Users need to be selected according to their profile of characteristics and according to the areas, which need to be tested in the interface or website. According to Cato, for "observed testing trails, you need to carry out six individual test sessions with users to obtain meaningful and useful results. Recruit six users for think aloud tests, and twelve for co-participation" (Cato 2001, p. 196). These sessions should be "clearly focused, objective, fast, and cost-effective" (Cato 2001, p. 196). More users can be recruited for website testing by putting messages on appropriate bulletin boards or via a recruitment agency.

When recruiting users for involvement in participative design, it is best to use real users who are dealing with the interface (i.e., website) very frequently. On the other hand, if real users cannot be recruited, the designer needs to work with "surrogates" such as students from universities and colleges who have an interest in working closely with the interface (i.e., websites) and who are reasonably representative of actual users.

Besides the above, designers need to include:

- Members of the steering committee for the project.
- Members of [the] design team or workshops.
- Reviewers who access the user interface.
- Test users [for] usability tests,
- Test users who exercise the system at delivery time to check that everything works correctly; and
- "Knowledge sources of how task and business procedures are currently carried out" (Lauesen 2005, p. 474).

Preferably, the designer should work very closely with the users to understand why they will use the website and to know exactly how and why particular tasks occur (and in what sequence), the types of problems that are facing the users, and the reasons for these. The designer needs to keep in mind that neither the manager nor the developer will be the type of users working with this website (or system), as both of them are in a different category from the users who are dealing with the website as part of their day-to-day work.

Users who are not in the expert category need support and help (i.e., documentation) from the developer to know how to work with this website (or system) to achieve their goals. Help and support are very important to the users, as via this information, the users can figure out which steps are needed to carry out their task. Therefore, documentation should contain clear, sequential steps in the correct order to allow the users to work efficiently to achieve the target.

3.3.4 Techniques for Observing and Listening to Users

Users are the main source of information for developing an interface such as a website. Therefore, a designer needs to acquire this information to develop and build a website. According to McCracken and Wolfe (2004, p. 44), there are a few golden rules which need to be taken into consideration from the designer's perspective, which include listening to users, "preferably in the context of the place where they will use your website"; and talking to the people who "use your website as part of the work they do on the job and to users who access your website without assistance or interaction with others, at home or work."

In this section, several techniques are discussed that will help the designer to gather more information about the users and their tasks. McCracken and Wolfe (2004, p. 49) states, "Users are in the business of doing their jobs, not explaining how they do their jobs, so simply asking 'How do you do your job?' will not give you the insights you need." Hence, appropriate techniques must be used in order to obtain information from users in an efficient and effective manner. Among the available techniques are: Interviews; Questionnaires; Think Aloud; Talk Right After; Protocol Analysis; Focus Group; and Mailed Surveys. They may be described as follows:

- *Interviews*: Set questions should be asked the users to gain more information about the system. Usually, the interviews occur face to face or via telephone. The purpose behind using this technique is to "gain information about a system and how it is, or will be used" (Bonharme 1996). Generally, three types of interviews can be used:

  - *Unstructured*: are not directed by a script; data, it is rich but not replicable.
  - *Structured*: are tightly scripted, often like a questionnaire. Replicable but may lack richness.

- *Semi-structured*: combine features of structured and unstructured interviews and use both closed and open questions. (Preece et al. 2002)

- *Questionnaires*: Collecting users' subjective opinions about a system can remove unpopular and unusable parts early in the design or after delivery. While interviews provide qualitative data, surveys and questionnaires provide quantitative data which can be statistically analyzed" (Bonharme 1996). Generally, two types of questions can be used—open or closed.

  - *Open Questions*: the user is free to provide his/her own answer; however, open questions are difficult to analyze in any rigorous way, or to compare, and can only be viewed as supplementary (Dix et al. 1993, p. 433).
  - *Closed Questions*: the user is asked to select an answer from a choice of alternative responses. For example, "there are several rating scales to choose from including, 3-point (yes/no/don't know), ranked order (numbering the options in order of preference), and bi-polar (good/bad)" (Bonharme 1996).

- *Think Aloud*: This technique is very simple and easy to use. It involves asking users to comment on their activities and aspects of the interface while working. This technique was developed by Erikson and Simon for investigating people's problem-solving strategies and is known as "cooperative evaluation as the user sees himself/herself as a collaborator in the evaluation and not simply as an experimental subject" (Dix et al. 1998, p. 427). This technique requires people "to say out loud everything that they are thinking and trying to do, so that their thought processes are externalized" (Preece et al. 2002, p. 365). The role of the designer is very important as he/she tries to keep the users talking while they are working at their task, whatever that task is, be it simple or difficult. The most important aspect of this technique is to listen very careful to the users discussing the work, their experience, and the environment in which they work. One drawback of this technique is that "thinking aloud" consumes some of the users' cognitive capacity and hence may inhibit their use of the system, biasing the results.

- *Talk Right After*: This technique can be used as an alternative to "Think Aloud" technique as some users cannot speak to the designer while they are working, for example a "travel agent, who is helping someone with questions, can't [cannot] speak to the designer and the customers simultaneously" (McCracken and Wolfe 2004, p. 50). Therefore, to prevent any disruption to the user's performance of the task, the designer can take notes about the tasks and later he/she can discuss it with the user.

- *Protocol Recoding*: There are a number of methods and techniques for recording user actions, for example:

  - *Paper and Pencil*: This is a low-technology technique, but a cheap and simple method for collection information from the user. This method "will allow the designer to note interpretations and extraneous events as they occur. However, this method has limitations in obtaining "detailed information as it is limited to the analyst's writing speed" (Dix et al. 1998, p. 428).

- *Audio and Video Recording*: In this technique, the user will be taped during his/her work, and later, the designer will study this tape and take notes of the user's activities. Therefore, this technique is very sensitive and responsive, so the user should be informed in this case, to avoid ethical problems.
- *Computer Logging*: is to get the system "automatically to record user actions at a keystroke level" (Dix et al. 1998, p. 428).

• *Focus Group*: This technique is very common in marketing, political campaigning, and social science research. In this technique, a small number of people (between 5 and 10 users) gather together to discuss a number of prepared questions. A mediator runs the meeting. The most important issue is that actual users should be involved in this step to provide more information and to bring consideration of real problems into the discussion. Normally, the session runs for an hour to an hour and a half.

 - The *advantages* of using this technique are:
    Focus group is low cost and easy to do. In addition, it provides quick results and is easy to scale to gather more data.
 - The disadvantages of working with this technique are:
    Serious problems can occur if one or two people dominate the entire discussion; therefore, the information will be gathered only from two instead of all the users (Preece et al. 2002). Therefore, an "effective facilitator will attempt to draw everyone into the discussion but will not always be successful" (McCracken and Wolfe 2004, p. 51)

• *Mailed Surveys*: This technique is cheaper for distribution to the users who are dealing with the interface. However, a lot of disadvantages can occur while working with this technique, for example (Fink 2012, 2010; Lesser et al. 2011):

 - Takes a lot of skill to write questionnaires that will obtain the information you want.
 - Some groups may interpret the questionnaires in their own way, and this will affect the results at the end.
 - Very few people respond to the mailed survey, and this will affect the results

• *Web Surveys*: are "powerful tools for maintaining respondent interest in the survey and for encouraging completion of the instrument" (Couper et al. 2001, p. 251). This technique is self-administered and involves computer-to-computer communication over the Internet, by asking the users to respond to the survey by clicking on radio buttons and adding additional comments in a specific area within the survey regarding the survey questions. Couper et al. (2001, p. 246) states, "Radio buttons are preferred because this allows mouse-only entry. In addition, radio button version would take less time to complete than the entry box version, given the added burden of typing numbers versus clicking a button." Web surveys are cost savings, speedy, offers greater anonymity, convenience and more sustainable compared with the previous techniques since they are designed and aimed to provide a more dynamic interaction between respondent and questionnaire compared with the paper mail survey. However, online surveys have disadvantages such as technical failures, computer viruses, Internet crimes, and hacking into the Web-based survey; these aspects can lead to a decrease in the response rate (Dillman 2007; Issa 2013).

- *Field Study*: Field studies are "done in natural settings with the aim of increasing understanding about what users do naturally and how technology impacts them" (Preece et al. 2002, p. 342). Field studies help the designers to identify opportunities for new technology, determine requirements for design, facilitate the introduction of technology, and evaluate technology. Furthermore, field studies get the team "immersed in the environment of their users and allow them to observe critical details for which there is no other way of discovering" (Spool 1997).

The designer must carefully consider the data requirements before an interview (or other data gathering technique) is conducted with the users. The designer needs to address the following issues before the interview:

- Understanding the concepts behind the interface.
- Defining the issues, which need to be clarified from the user such as—tasks, problems, and procedures, which need to be followed to accomplish a specific task.

Throughout the above stages, the designer will gather some information about the interface itself, the tasks, problems, and the steps to accomplish the tasks. If the information does not meet their requirements, then it may be better to apply an alternative information gathering technique before moving to the next step in the methodology.

3.3.5   Internet Marketing and User Responses

There are other ways of determining website users' needs and desires. Internet marketing is a new approach, where customers can define "what information they need, what offering they are interested in, and what price they are willing to pay" (Sheth et al. 2001, p. 6).

According to Hoffman and Novak (1996, p. 51), the Internet is an important focus for marketers for several reasons:

- Consumers and firms are conducting a substantial and rapidly increasing amount of business on the Internet.
- The market prefers the decentralized, many-to-many web for electronic commerce to the centralized, closed-access environments provided by the online services.

- The World Wide Web represents the broader context within which other hypermedia Computer-Mediated Environment (CMEs) exit.
- The web provides an efficient channel for advertising, marketing, and even direct distribution of certain goods and information services.

Consequently, Internet marketing is using the Internet and web as a medium to provide information to customers globally. Since it changes rapidly, with new tools being developed to attract more customers to use it, it is important to establish the requirements for interactive marketing. This depends on three issues—"direct communication, individual choice, and friendly technology" (Hanson 2000, p. 95). These address the requirements by learning about each customer's attitudes and behaviors.

On the Internet, several tools can be used by the user to gain more information about specific products or by asking the user to give some feedback about the products. Examples of these tools are user response form, forums, and chat rooms. These tools have two advantages: (1) They encourage the user to provide feedback about the website layout or asking questions about the products in general; (2) they reduce the web master's job by posting all the answers in one place, thereby allowing the users to check the answers from one place.

- *User response form*: this type will allow the user to enter his/her message or checking some fields "can vary from checkbox type responses to the provision of text areas" (Darlington 2005, p. 65). Some systems will be capturing the data from the user response and sending the answer to the user via the e-mail.
- *Forums*: are called "bulletin boards" or "newsgroups"; this type of facility provides discussion forums for people with similar interests. For example, "they can also serve as a source of feedback as someone can start a discussion by posting comments about a subject another person may answer, to be followed by other people joining and so on, so a thread of linked messages develops" (Darlington 2005, p. 66).
- *Chat rooms*: are called Internet relay chat {IRC} channels and "allow groups of people to exchange live text messages" (Darlington 2005, p. 67).
- *Blogs*: are called "Web log" or "blogging"; this type of facility has the ability to create an online text diary, "made up of chronological entries that comment on everything from one's everyday life to wine and food to computer problems" (Jessup and Valacich 2008, p. 210). This facility can give an easy method of "publishing web pages which can be described as online journals, diaries or news or events listings" (Chaffey 2007, p. 99).

3.4 Conclusion

This chapter discoursed and studied user participation in the system development process, since it is essential to involve users in the design stage to reduce the gap between users and designers' goals and users and computers on the other.

Currently, there are various types of devices in the market, i.e., software applications, mobile, and portable devices (e.g., iPads, iPhone) but the majority of these
devices are still poorly designed and user satisfaction is inadequate. This chapter
presented and addressed user participation significance in the design process by
discussing several sections in relation how we know our users, recruiting users and
managing user participation in the development processes.

User participation is essential in the sustainable design as well as to improve
device acceptance among the users and satisfy their needs. Finally, user participation
is vital and fundamental in the system development process along with sustainable
design to increase users' acceptance and satisfaction.

References

Amoako-Gyampah K, White KB (1993) User involvement and user satisfaction: an exploratory
     contingency model. Inf Manag 25(1):1
Applen JD (2002) Tacit knowledge, knowledge management, and active user participation in website
     navigation. IEEE Trans Prof Commun 45(4):302
Barki H, Hartwick J (1989) Rethinking the concept of user involvement. MIS Q 13(1):53
Baroudi JJ, Olson MH, Ives B (1986) An empirical study of the impact of user involvement on
     system usage and information satisfaction. Commun ACM 29(5):232–238
Bonharme E (1996) Usability evaluation techniques. http://www.dcs.napier.ac.uk/marble/Usability/
     Evaluation.html. Accessed 25 Oct 2021
Cato J (2001) User-centered web design. Pearson Education Limited
Chaffey D (2007) E-Business and E-Commerce management. Pearson Education Limited, England
Couper MP, Traugott MW, Lamias MJ (2001) Web survey design and administration. Public Opinion
     Q 65(2):230–253
Darlington K (2005) Effective website development : tools and techniques. Pearson Education
     Limited 2005, UK
Dean DL, Lee JD, Pendergast MO, JFN Hickey AM Jr (1997) Enabling the effective involvement
     of multiple users: methods and tools for collaborative software engineering. J Manag Inf Syst
     14(3):179
Dillman D (2007) Mail and internet surveys the tailored design method, 2nd edn. John Wiley &
     Sons Inc., USA
Dix A, Finlay J, Abowd G, Beale R (1993) Human computer interaction. Pearson Prentice Hall,
     England
Dix A, Finlay J, Abowd G, Beale R (1998) Human-computer interaction, 2nd edn. Pearson Education
     Limited, USA
Doll WJ (1987) Encouraging user management participation in systems design. Inf Manag 13(1):25
Doll WJ, Torkzadeh G (1989) A discrepancy model of end-user computing involvement. Manage
     Sci 35(10):1151
Doll WJ, Torkzadeh G (1991) A congruence construct of user involvement. Decis Sci 22(2):443
Engler N (1996) Bringing in the users. Computerworld 30(48):3
Fink A (2010) Survey research methods. Educ Res Methodol Quant Methods Res:152-160
Fink A (2012) How to conduct surveys: a step-by-step guide. Sage Publications, California, USA
Fleming J (1998) Web navigation: designing the user experience. O'Reilly & Associates, CA, USA
Hackos JT, Redish JC (1998) User and task analysis for interface design. Wiley Computer Publishing
Hanson W (2000) Principles of internet marketing. South-Western College Publishing, Thomson
     Learning, Ohio, USA

Hartwick J, Barki H (1994) Explaining the role of user participation in information systems. Manage Sci 40(4):440

Hawk SR, Dos Santos BL (1991) Successful system development: the effect of situational factors on alternate user roles. IEEE Trans Eng Manage 38(4):316

Hepworth JB, Vidgen GA, Griffin E, Woodward AM (1992) The enhancement of information systems through user involvement in system design. Int J Inf Manage 12(2):120

Hirschheim R (1985) User experience with and assessment of participative systems design. MIS Q 9(4):295

Hoffman DL, Novak TP (1996) Marketing in hypermedia computer-mediated environments: conceptual foundations. J Market 60:50–68

Issa T (2013) Online survey: best practice. In: Isaias P, Nunes MB (eds) Information systems research and exploring social artifacts: approaches and methodologies. IGI Global, USA, pp 1–19. https://doi.org/10.4018/978-1-4666-2491-7.ch001

Jessup L, Valacich J (2008) Information systems today: managing in the digital world. Pearson Education Limited, USA

Jiang JJ, Chen E, Klein G (2002) The importance of building a foundation for user involvement in information system projects. Project Manag J 33(1):20

Kappelman L (1995) Measuring user involvement: a diffusion of innovation perspective. Data Base Adv Inf Syst 26(2/3):65–86

Kawalek P, Wood-Harper T (2002) The finding of thorns: user participation in enterprise system implementation. Data Base Adv Inf Syst 33(1):13–22

Lauesen S (2005) User interface design a software engineering perspective. Pearson Education Limited,

Lawrence M, Low G (1993) Exploring individual user satisfaction within user-led development. MIS Q 17(2):195

Lesser VM, Yang DK, Newton LD (2011) Assessing hunters' opinions based on a mail and a mixed-mode survey. Human Dimensions Wildlife 16(3):164–173. https://doi.org/10.1080/10871209.2011.542554

Lin WT, Shao BBM (2000) The relationship between user participation and system success: a simultaneous contingency approach. Inf Manag 37(6):283–295. https://doi.org/10.1016/S0378-7206(99)00055-5

McCracken DD, Wolfe RJ (2004) User-centered website development: a human-computer interaction approach. Pearson Education Inc., New Jersey, USA

McKeen JD, Guimaraes T, Wetherbe JC (1994) The relationship between user participation and user satisfaction: an investigation of four contingency factors. MIS Q 18(4):427

Merrick B (2001) Eight steps to achieve e-commerce success. Credit Union Mag 67(2):13

Mumford E (1995) Effective systems design and requirements analysis. Macmillan Press Ltd., Great Britain

Olson M, Ives B (1981) User involvement in system design: an empirical test of alternative approaches. Inf Manag 4(4):183

Preece J, Rogers Y, Benyon D, Holland S, Carey T (1994) Human computer interaction. Addison-Wesley, England

Preece J, Rogers Y, Sharp H (2002) Interaction design: beyond human-computer interaction. John Wiley & Sons, USA

Rondeau PJ, Vonderembse MA, Ragu-Nathan TS (2002) Investigating the level of end-user development and involvement among time-based competitors. Decis Sci 33(1):149

Saleem N (1996) An empirical test of the contingency approach to user participation in information systems development. J Manag Inf Syst 13(1):145

Sheridan W (1999) Web design is changing. http://www3.sympatico.ca/cypher/web-design.htm. Accessed 25 Oct 2021

Sheth JN, Eshghi A, Krishnan BC (2001) Internet marketing. Harcourt, Orlando, USA

Spool JM (1997) Field studies: the best tool to discover user needs. http://www.uiconf.com/8/articles/field_studies_article.html. Accessed 25 Oct 2021

Tait P, Vessey I (1988) The effect of user involvement on system success: a conting. MIS Q 12(1):91
Turk A (2001) Towards contingent usability evaluation of WWW sites. In: Proceedings of OZCHI, Perth, Australia, pp 161–167

Chapter 4
Physical, Cognitive, and Affective Engineering

Abstract This chapter will assess the importance of physical, cognitive, and affective engineering in designing and developing technology for the user interface, device, and website. These aspects are essential in technology design, since they assist designers to examine the relationships between users and technology, and to improve users' performance when dealing with this technology, in order to reduce errors and increase satisfaction and users' acceptance of the system. These aspects should be part of sustainable technology design to ensure the users' acquiescence, reduce their frustration, and ensure that the new smart technology design will meet user, society, and community needs simultaneously.

4.1 Introduction

In this chapter, the authors explore the physical, cognitive, and affective aspects of engineering. Physical engineering examines how users' physical abilities will interact with and affect the ways in which users perform tasks using technology; cognitive engineering applies knowledge of cognitive attitude in the development of interactive systems. Finally, affective engineering explains how and why users cooperate with technology and how this can be applied to design. This chapter provides to designers and users clear guidelines regarding these concepts and indicate how and why these concepts are essential in technology design; furthermore, it explains how designers can measure and evaluate physical, cognitive, and affective engineering features in terms of users' requirements.

This chapter is organized as follows: physical engineering, cognitive engineering, Goals, Operators, Methods, and Selection Rules (GOMS), Norman's model, and affective engineering.

© Springer-Verlag London Ltd., part of Springer Nature 2022
T. Issa and P. Isaias, *Sustainable Design*,
https://doi.org/10.1007/978-1-4471-7513-1_4

4.2 Physical Engineering

This study aims to combine human body mechanics and physical limitations with industrial psychology to facilitate the interaction between human and devices in order to improve people's job performance and cater for users' needs.

Physical engineering aims to improve users' performance ability by handling the workload in the workplace, as improved performance is concerned with reducing errors, improving quality, reducing the time required to complete tasks, and ensuring and ascertaining users' acceptance of the system (see Fig. 4.1).

The physical engineering aspects of human–computer interaction come into play principally in the process of input and output devices. The main objective of using input devices is to control the system's operations and input data, an example of input devices, mouse, joystick, text, numeric, graphic data, drawing, voice, and touch. On the other hand, output devices are machines used to represent data from other devices, i.e., monitors, printers, auditory output, synthesized speech, visual display, wearable devices, wireless devices, and haptic devices.

Physical engineering is also concerned with the ergonomics of information systems. It is concerned with things such as the physical workstation and furniture design, lighting, noise, and keyboard height and arrangement. These are all physical aspects of human engineering within an information systems context.

Currently, devices in general are being increasingly used to assist people to improve their job and work performance and productivity. This includes individuals with hearing, vision, or other physical impairment(s). Designers of new smart technologies should consider ways by which to improve the quality of life of people with disabilities and encourage them to be part of the society and community.

Fig. 4.1 Physical engineering—prepared by Tomayess Issa

A well-designed computer interface must take into consideration human limitations since those with disabilities must be considered as members of the community and society in general. Therefore, HCI experts and designers must include these categories of people in their agenda in order to serve them and provide the necessary facilities allowing them to become self-determining and independent. Examples of physical human limitations include (Te'eni et al. 2007; Zhang et al. 2005):

- *Sensory limit*: what and how much our senses can perceive.
- *Responder limit*: reach and strength.
- *Cognitive limit*: reaction time and accuracy.
- *Other limitations*: vision, audition, touch, and motor-related activities.

Furthermore, HCI experts and designers should provide the necessary guidelines and principles for accessibility, especially in the new smart technology devices. These guidelines and principles are specified in Dix et al. (1993), Gerlach and Kuo (1991), Issa and Turk (2010), and Te'eni et al. (2007):

- *Standardize Task Sequences*: allow users to perform tasks in the same sequence and manner across similar conditions.
- *Ensure the embedded links are descriptive*: using embedded links, the links text should accurately describe the link's destination.
- *Use unique and descriptive headings*: use headings that are different from one another and conceptually related to the content they describe.
- *Use radio buttons for mutually exclusive choices*: provide a radio button control when users need to choose one response from a list of equally exclusive options.
- *Non-text Element*: provide a text equivalent for every non-text element.
- *Synchronize*: for any time-based multimedia presentation synchronize equivalent alternatives.
- *Color*: information conveyed with color should also be conveyed without it.
- *Title*: title each frame to facilitate identification and navigation.

Furthermore, Smith and Mosier (1986) offer five high-level goals for designing user interface software including the new smart technology and devices for human beings in general:

- *Consistency of data display*: Formats, colors, capitalization, and so on should all be standardized and controlled by use of a dictionary of these items.
- *Efficient information assimilation by the user*: Format should be familiar to the user and should be related to the tasks required to be performed with the data.
- *Minimal memory load on the user*: Users should not be required to remember information from one screen for use on another screen.
- *Compatibility of data display with data entry*: The format of displayed information should be linked clearly to the format of the data entry.
- *Flexibility for user control of data display*: Users should be able to obtain the information from the display in the form most convenient for the task on which they are working.

Furthermore, Shneiderman and Plaisant (2010) establish several guidelines for HCI experts and designers so that their technology design engages users' attention by effectively using features such as intensity, marking, size, fonts, video, blinking, color, and audio.

- *Intensity*: use two levels only, with limited use of high intensity to draw attention.
- *Marking*: underline the item; enclose it in a box; point to it with an arrow.
- *Size*: use up to four sizes to draw attention.
- *Fonts*: use up to three fonts.
- *Video*: use opposite coloring.
- *Blinking*: use blinking displays or blinking color changes with great care and in limited areas.
- *Color*: use up to four standard colors, with additional colors reserved for occasional use.
- *Audio*: use soft tones for regular positive feedback and harsh sounds for rare emergency conditions.

Therefore, HCI experts and designers should adopt these guidelines in their agenda and design technologies in order to minimize user frustrations and obstructions and to support disabled people who use devices ranging from workstations to new smart technologies such as iPads or iPhones.

Furthermore, to measure physical engineering, designers must measure safety, audible, and readable. By following these measurements, designers will ensure that the new smart technology meets users' requirements (Card et al. 1983; Preece et al. 1994; Shneiderman 1986).

Finally, several studies (Card et al. 1983; DePaula 2003; Dix et al. 1993; Gerlach and Kuo 1991; Olson and Olson 2003; Preece et al. 1994; Te'eni et al. 2007) indicate that technology and devices are being used more and more to assist users and disabled individuals to accomplish tasks; however, this technology can cause major health risks involving vision and muscular problems, and this can lead to inflammation, disk problems, and painful muscles. Therefore, designers should initiate an awareness campaign for the new generation (called Internet generation), since these people depend to a great extent on technology for their study and work. This awareness should be available on various media including websites, Facebook, and the devices' packaging.

4.3 Cognitive Engineering

Cognitive processes involve user activities including thinking, reading, writing, talking, remembering, making decision, planning, solving problems, and understanding people (see Fig. 4.2). Norman (1993) distinguishes two types of cognition, namely: experiential and reflective. The Experiential mode reflects perceive, act, and react, as it needs a certain level of motivation and enthusiasm, i.e., driving a car, reading a book, playing a video game, or having a conversation. On the other hand,

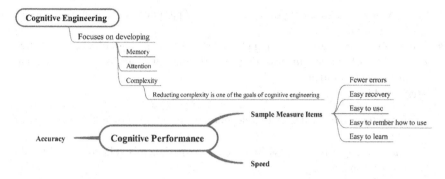

Fig. 4.2 Cognitive engineering—prepared by Tomayess Issa

the reflective mode involves thinking, comparing, and decision making. This mode leads to creativity and innovation such as writing a book, designing, and learning (Isaias and Issa 2015).

Overall, both modes need specific technologies and are essential for everyday life.

Cognitive engineering focuses on developing systems which support cognitive processes of users such as memory, perception and recognition, memory, learning, reading, speaking, listening, problem solving, decision making, and attention that are used in HCI. Difficulty is seen to represent the employment of rare cognitive resources, and reducing complication is one of the goals of cognitive engineering (Isaias and Issa 2015).

The human information processing [HIP] model validates how cognitive resources such as memory and processors are employed. There are three types of processors, namely—[1] perceptual: detects and accepts inputs from the external world and stores parts of the input in the working memory; [2] cognitive: interprets, manipulates, and makes decisions about the inputs; and [3] motor: is responsible for translating cognitive decisions into physical actions such as using a keyboard. There are two types of memory, namely: working memory which is similar to the human brain's task, since information and data are coming to the human brain for processing and storage of complex cognitive tasks such as language, learning, comprehension, and reasoning (Baddeley 1992); and long-term memory which permanently stores, manages, and retrieves information for future use and life time (Goelet et al. 1986).

Generally, cognitive engineering takes a narrow view in relation to performance, automatic behavior, controlled behavior, processing of images, processing of verbal information, and memory aids (Te'eni et al. 2007, pp. 89–90).

- *Performance*: the speed and accuracy of the information processing task.
- *Automatic behavior*: fast and relatively undemanding of cognitive resources (i.e., entering 50 numbers into a spreadsheet would quickly become an automatic activity).
- *Controlled behavior*: slow and cognitively demanding (i.e., deciding to use the summation function and defining it parameters requires access to long-term

memory, selection of appropriate functions and parameters and control to ensure correct operation).

- *Processing of images*: processing characterized as spatial, graphic, and holistic.
- *Processing of verbal information*: processing characterized as sequential, linguistic, and technical.
- *Memory Aids*

  - *Heuristics*: rules of thumb that depend heavily on the content and context of the task.
  - *Image*: a cognitive process in which an experience is related to an already familiar concept.
  - *Mental model*: a representation of the conceptual structure of a device or a system.

Cognitive engineering focuses on development systems that support and assist designers to understand the interaction between the user and the technology (including computer). Similarly, Gersh et al. (2005) indicate that cognitive engineering developed in response to two reasons: first, to ensure that technologies including computers are well designed and meet users' needs; secondly, it introduced design principles in technology design to ensure that skilled technicians could operate them safely and efficiently.

Finally, in order to measure cognitive engineering, designers should consider the following measures in technology design, namely: fewer errors, easy recovery, easy to use, easy to remember, how to use, easy to learn (Dix et al. 1998, 2004).

4.4 Goals, Operators, Methods, and Selection Rules (GOMS)

The Goals, Operators, Methods, and Selection Rules (GOMS) model was created by Card et al. (1983). This model aims to present the knowledge of determined human–computer interaction (HCI), and how users can interact with computers and the implications for designers. This model endeavors to reduce the complexity in the interface as well as in the cognitive resources and engineering. This model has specific elements that describe purposeful HCI:

- *Goals* specify what the user wants and intends to achieve.
- *Operators* are the building blocks for describing human–computer interaction at the concrete level.
- *Methods* are programs built with operators that are designed to accomplish goals.
- *Selection rules* predict which method will be used. For example, "If the mouse is working, select 'point to an item on screen', if not select 'choose OPEN option in file menu'."

Finally, the GOMS model (Goals, Operators, Methods, and Selection Rules) is based on levels of interaction that bridge the gap between the abstract (psychological) task and the concrete (physical system).

4.5 Norman's Model

To understand the interaction between human and computer, Norman developed a model of user activity (Norman 1986). Before discussing Norman's model, we need to understand the principles of human behavior in order to enhance users' performance in terms of an effective design and technology. These principles are divided into gulf of execution which handles the interruption between the user's goal and aims and its device implementation, and the gulf of evaluation that relates to the gap between device implementation of the user's goal and its evaluation by the user (Te'eni et al. 2007).

Norman's model has eight steps intended to assist users to complete and accomplish a task when using a specific technology:

- *Goals*: create a goal that needs to be accomplished.
- *Intentions*: develop an intention that will accomplish the goal.
- *Action Specification*: identify a sequence of actions to implement the intentions.
- *Execution*: execute the action.
- *Perception*: understand the system outcomes from the action.
- *Interpretation*: interpret the system state.
- *Evaluation*: evaluate the results and compare it with the goals.

Figure 4.3 shows the steps that jointly required the user goals for a particular goal. Generally, these steps will allow users to identify their goals: what is done to the world, the world, and to check the world. In general, these steps have three majors' components: identify the goals, do something, and evaluate at the end.

4.6 Affective Engineering

Affective engineering focuses mainly on emotions, moods, affective impressions, and attitudes; it concentrates on integrating product design and consumers' feelings for a product into design elements (Hewett et al. 1992; Jordan 2002; Rosson and Carroll 2001).

Affective engineering is essential in human–computer interaction to balance and integrate the affective and cognitive aspects in the technology design; cognitive engineering interprets and makes sense of the world, while affective engineering evaluates, judges, and provides some warning to the users out of possible hazards and risks.

Fig. 4.3 Norman's seven stages (adopted from Norman 1986)—prepared by Tomayess Issa

Affective engineering is used in any technology design ranging from user interface, technology or websites to color, animation, layout, structure, text, images, and menu. For example, using pastel colors for e-commerce sites will leave users feeling calm and will foster a more accepting attitude and readiness to buy and interact further with the site. Additionally, affective engineering focuses on technology design, which is pleasing, engaging, enjoyable, fun, attractive, beautiful satisfying, and entertaining. These attributes will encourage the user to accept and use the new smart technology to achieve his/her goals and aims (see Fig. 4.4).

Furthermore, user attitudes to combined cognitive and affective engineering are used to evaluate devices including computers, mobiles, and other devices. The evaluation aims to identify errors and problems in order to ascertain whether or not the devices are successful. This evaluation is based on users' perceptions and opinions and should be taken into account by designers in order to resolve any problems and meet user needs.

Attitudes can be shaped and managed to some extent by training users to examine the devices' performance in general in order to reduce anxiety. Furthermore, a very important step in the design process is the management and involvement of users, as this will promote user satisfaction and acceptance of devices, further reducing user frustration.

Finally, to ensure that users will accept devices, satisfaction is considered the most commonly used in the HCI and information systems field, since users will either confirm or not confirm their satisfaction with the device.

Fig. 4.4 Affective engineering and measured satisfaction—prepared by Tomayess Issa

Doll and Torkzadeh (1988) proposed the most popular measure of satisfaction called End-User Computer Satisfaction. This measure is constructed of five sub-factors, namely: content, accuracy, format, timeliness, and ease of use.

According to Doll and Torkzadeh (1988, p.268), the five sub-factors include the following aspects:

Content

- Does the system provide the precise information that the user needs?
- Does the information content meet user needs?
- Does the system provide reports that meet user needs?
- Does the system provide adequate information?

Accuracy

- Is the system accurate?
- Is the user satisfied with the system accuracy?

Format

- Is the system output presented in useful format?
- Is the information clear?

Ease to use

- Is the system user-friendly?
- Is the system easy to use?

Timeliness

- Does the system provide the information that you need in time?
- Does the system provide up-to-date information?

The End-User Computer Satisfaction instrument is a significant development, as it will assist designers to measure user satisfaction with a technology design. This evaluation and measurement will assist designers to identify any errors and problems in their design, making it easier for them to tackle these problems in order to improve users' satisfaction and acceptance.

4.7   Conclusion

This chapter discussed and examined several features, which are required for technology design including sustainable design. These include physical, cognitive, and affective engineering. Physical engineering is mainly concerned with the user's ability to handle the load or demands of the work situation, job performance (i.e., reduce errors, enhance quality, and reduce time required to complete specific tasks), and acceptance of the system. Cognitive engineering involves user activities including thinking, reading, writing, talking, remembering, making decision, planning, solving problem, and understanding people. This engineering is mainly intended to reduce the complexity between users and devices. Finally, effective engineering works alongside physical and cognitive engineering to examine and assess users' emotions, moods, impressions, and attitudes toward product design.

References

Baddeley A (1992) Working memory. Science 255(5044):556–559

Card S, Moran TP, Newell A (1983) The psychology of human computer interaction. Lawrence Erlbaum Associates Hillsdale, N.J.

DePaula R (2003) A new era in human computer interaction: the challenges of technology as a social proxy. In: Latin American Conference on HCI, Rio de Janeiro, Brazil 2003. ACM International Conference Proceeding series, pp 219–222

Dix A, Finlay J, Abowd G, Beale R (1998) Human-computer interaction, 2nd edn. Pearson Education Limited, USA

Dix A, Finlay J, Abowd G, Beale R (1993) Human computer interaction Pearson Prentice Hall England

Dix A, Finlay J, Abowd G, Beale R (2004) Human-computer interaction, 3rd edn. Pearson Education Limited, England

Doll WJ, Torkzadeh G (1988) The measurement of end-user computing satisfaction. MIS Q 12(2):259–274. https://doi.org/10.2307/248851

Gerlach JH, Kuo F-Y (1991) Understanding human computer interaction for information systems design. MIS Q 14(4):526–549

Gersh JR, McKneely JA, Remington RW (2005) Cognitive engineering: understanding human interaction with complex systems. Johns Hopkins APL Technical Digest 26(4):377–382

Goelet P, Castellucci VF, Schacher S, Kandel ER (1986) The long and the short of long-term memory: a molecular framework. Nature 322:419–422

Hewett TT, Baecker R, Card S, Carey T, Gasen J, Mantei M, Perlman G, Strong G, Verplank W (1992) ACM SIGCHI curricula for human-computer interaction. ACM, USA

Isaias P, Issa T (2015) High level models and methodologies for information systems. Springer, New York

Issa T, Turk A (2010) Usability and human computer interaction in developing websites: an Australian perspective. Paper presented at the In Proceedings of the IADIS International conferences: interfaces and human computer interaction 2010 and game and entertainment technologies 2010, Freiburg, Germany

Jordan PW (2002) Designing pleasurable products: an introduction to the new human factors. CRC Press, Great Britain

Norman DA (1986) Seven-Stage model of (Individual) Interaction. Dept of Computer and Systems Sciences, Lulea University, Bai Guohua, Sweden

Norman D (1993) Things that Make Us Smart. In. Addison-Wesley, USA,

Olson G, Olson J (2003) Human computer interaction: psychological aspects of the human use of computing. Annu Rev Psychol 54:491–516

Preece J, Rogers Y, Benyon D, Holland S, Carey T (1994) Human computer interaction. Addison-Wesley, England

Rosson MB, Carroll JM (2001) Usability engineering: scenario-based development of human-computer interaction. Elsevier, CA, USA

Shneiderman B, Plaisant C (2010) Designing the user interface: strategies for effective human-computer interaction. Addison Wesley, Boston, MA, USA

Shneiderman B (1986) Designing the user interface-strategies for effective human-computer interaction. Pearson Education India, Boston, MA, USA

Smith SL, Mosier JN (1986) Guidelines for designing user interface software. Mitre Corporation Bedford, MA, Bedford, Massachusetts, USA

Te'eni D, Carey J, Zhang P (2007) Human computer interaction: developing effective organizational information systems. John Wiley & Sons Inc., USA

Zhang P, Carey J, Te'eni D, Tremaine M (2005) Integrating human-computer interaction development into the systems development life cycle: a methodology. Commun Assoc Inf Syst 15:512–543

Chapter 5
Color, Prototyping and Navigation, Principles and Guidelines Design, Evaluation and Testing, and Task Analysis

Abstract This chapter discusses the color, prototyping and navigation, principles and guidelines design, evaluation and testing, and task analysis pertaining to the new smart technology design. These are vital aspects of design that must be taken into account by the designers and HCI experts, by integrating these aspects in the new smart technology design, the new device, user interface, and website that will meet the needs of users, the community, and society in general. Therefore, if all these design considerations are taken into account, users will have full control of their devices without any frustration and irritation as users have the opportunity to evaluate and test them in order to meet their needs. Moreover, designers and HCI experts should consider these aspects in their new smart technology design to ensure their new design is in accordance with sustainability principles.

5.1 Introduction

To ensure that new smart technology design is widely accepted and used effectively both globally and locally, designers should consider the following: color, prototyping and navigation, principles and guidelines design, evaluation and testing, and task analysis. These aspects are essential in any new smart technology design for devices, user interfaces, or websites. Users are becoming more sophisticated, and their expectations and behaviors concerning new smart technology design are changing as they have the autonomy to select a new smart technology design, which matches their needs. Therefore, HCI experts should consider the needs of users, the community, and society in order to ensure that the new smart technology design is designed based on sound design principles, which include the notion of sustainability. This chapter is organized as follows: color, prototyping and navigation, principles and guidelines design, evaluation and testing, and task analysis.

© Springer-Verlag London Ltd., part of Springer Nature 2022   75
T. Issa and P. Isaias, *Sustainable Design*,
https://doi.org/10.1007/978-1-4471-7513-1_5

5.2 Color

The consideration of color in a new smart technology design is vital as it can determine the success of failure of a device, interface, or website. Up until now, designers and HCI experts have used color based on their own individual, personal preferences rather than on scientific evidence (Holtze 2006, p. 34).

This approach will affect users' attitudes to these technologies in terms of style, layout, structure, navigation, usability and ad speed, and their acceptance or rejection of this new smart technology. Shneiderman and Plaisant (2010) posited that designers should limit the number of colors used in their designs and should select the colors which are the most appropriate for the contents and audience. Furthermore, Te'eni et al. (2007) verified that color usage in new smart technology design will help the user to understand and absorb information when reading, decision making, and differentiating between important and unimportant information.

Color theory in new smart technology design is considered in consumer-oriented websites that match the social and emotional perceptions of users and are expected to "increase trust and be more engaging, also increase user enjoyment or loyalty" (Cyr et al. 2010, p. 2). Color has played an important part in communication, psychology, and even physical health. Arguably, color has power, which is utilized for interior design, graphic design (Web or Interface), and art.

Generally speaking, color comprises three variables: hue, saturation (or chroma), and brightness (or intensity or luminance) (Holtze 2006; Pelet et al. 2013).

- **Hue**:

  - Corresponds to the normal meaning of color—changes in wavelength (these are spectral colors).

- **Saturation (or Chroma)**

  - Is the relative amount of pure light that must be mixed with the white light to produce the perceived color.

- **Brightness (or intensity or luminance)**

  - Refers to the shades of gray decreasing from white through gray to black.

There are three-color wheels, namely primary, secondary, and tertiary hues (Morton 2015)—primary hues: blue, red, yellow (in the printing world, these colors are cyan, magenta, yellow); secondary hues: violet, green, orange; and tertiary hues: red-violet, yellow-orange, blue-green, red-orange, blue-violet, yellow-green.

The judicious use of color in a new smart technology device has several advantages including: attracting attention, being appealing, facilitating recognition, and assisting memory and comprehension. Moreover, the choice of colors can help users to understand and recall information when undertaking reading and decision-making tasks, and supports effective processes, i.e., attract attention, help users to memorize, and add reminders.

There are two general design guidelines for color: Firstly, allow for redundancy so that differentiation by color is also accompanied by differentiation by shape or size. Secondly, whenever possible, authorize the users to adapt colors to suit their preferences and their culture.

Let us explain the effects and moods of color usage in new smart technology design. There are various types of colors from cold, cool, hot, warmth, darkness, light, pastel, "intensity" (power and passion) (Elliot and Maier 2012; Labrecque and Milne 2012; QSX Software Group 2015; Sibagraphics 2015).

- *'Cold' colors*:

  - Colors like blue, green, and blue-green are associated with coldness and calm.
  - Use these colors to promote a feeling of seriousness, significance, honesty, determination, cleanliness, refreshing freshness, coldness.

- *'Cool' colors*:

  - Blue is the base for these colors but added are reds and yellows to bring out a wide range of color from minty green to a soft violet.
  - These colors help promote a feeling of calm, serenity, trust, and relaxation.

- *'Hot' colors*:

  - Red is the highest chroma color, and is the most powerful hue.
  - A hot color may evoke strong emotional responses and has been known to stimulate physical activity and sexual desire.
  - Use hot colors if you want an aggressive feel or want something stand out among others.
  - Red is the strongest of hues, placing a high chroma yellow in any designed, or work of art will draw the eye first.

- *'Warm' colors*:

  - Based in red but softened and suffused with orange and yellows. Warm colors are often used to suggest comfort and warm, and heartfelt emotions.

- *'Darkness' colors*:

  - Black is a mysterious color associated with fear and the unknown.
  - They are often used to reduce space.
  - These colors are also used so that lighter colors can stand out greater and be more effective.
  - These colors are serious and can suggest depressed and hardness.

- *'Light' colors*:

  - These colors are barely colors at all; they exist merely as suggestions and hints of colors.

- They are the opposite of darkness, and they are often used to open up a space or evoke a feeling of openness.

- *'Pastel' colors*:

  - These pale colors are hues tinted with large amounts of white and are very soft in nature.
  - This type of color suggests innocence, fond memories, and romance.

- *'Intensity (power and passion)' colors*:

  - The colors of intensity are high chroma colors and pure and seem to scream their message. Great for attention grabbing.

In conclusion, several studies (Cornforth 1994; Morton 2010; Wang et al. 2008) indicate color is essential in new smart technology design as it can enhance marketing, especially in the brand recognition. Compared with black and white, the use of color will increase users' participation and engagement, especially in traditional (i.e., newspapers) and online facilities.

In general, using color in new smart technology design will attract attention, help users to memorize, and add reminders. Moreover, another powerful effect is that it facilitates recognition and comprehension by both the designers and the users.

5.3   Navigation

Navigation is concerned with finding out about, moving through, and the environment. It includes three different but related activities: object identification, which is concerned with understanding and classifying the objects in an environment, exploration, which is concerned with finding out about a local environment and how that environment relates to other environments, and wayfinding, which is concerned with navigation toward a known destination (Adler and Blue 1998; Elfes 1987; Taylor and Sennott 1984).

Furthermore, several studies (Blackmon et al. 2002; Fons et al. 2003; Kakumani et al. 2004) indicate that a part of navigation is labeling, as labels are used for internal and external links, headings, subheading, titles, and related areas. For example, there is nothing more confusing for people than a website changing its own vocabulary by referring, for example, to "products" one minute and "items" the next. The same labels should be used consistently on searching mechanisms and on the main pages, in the names of the pages and in the link names.

This type of job will assist the navigation support in any new smart technology design, as many of the signs and labels are deliberately placed in order to support navigation, and it is common to have a navigation bar across the top of a design (i.e., site) which points to the main, top-level categories. This is often called the "global navigation bar."

Within each of these, there will be sub-categories; these might be placed down the left-hand side of the site or may drop down when the main category is selected. This is known as "local navigation."

It is a good design principle to have the same global, top-level navigation bar on every page so that people can easily jump back to the home page, to a "frequently asked questions" page or to one of the other main categories.

An essential aspect of the navigation features of any new smart technology design is to provide a "YOU are here" sign. This is often presented by a description showing where people are in the hierarchy of the site. Other devices such as indexes and glossaries are helpful in assisting people find exactly what they are searching for. The site map should be made available so that it can be called up when needed.

One of the significant features of the new smart technology design as an information space is that many sites support the searching process. Search engines can be bought; the better ones are quite expensive but are also effective. Two main problems with searching a website are: The first is knowing exactly what sort of documents the search engine is searching for; the second is how to express a combination of search criteria.

- *Inclusion and Exclusion*

    - With many search engines, you can improve search performance by specifying an *"inclusion operator,"* which is generally a plus (+) sign. This operator states that you do not want a page retrieved unless it contains the specified word. By listing several key terms with this search operator, you can exclude many pages that do not contain one or more of the essential terms. The following, for example, will retrieve only those pages that contain all three of the words mentioned

        i.e., kittens + care + Siamese

- *Wildcards*

    - An asterisk* is a wild card.
    - That is, searching for hunt* will return sites with hunter, hunters, hunting, huntsman, etc.

- *Boolean Searches*

    - Use keywords (AND, OR, and NOT) to link the words you are searching for.

- By using Boolean operators, you can gain a more precise control over your searches.

    That is, AND operator tells the search service to return only those documents that contain both words

- That is, kittens AND care

    That is, OR operator is used to search for documents containing either word
- That is, Kittens OR care

    That is, NOT operator tells the search engine to omit any documents containing the word preceded by NOT (just as the minus sign does). For example, the search phrase "kittens NOT cats" retrieves pages that mention kittens but not those that mention cats.

- *Using Parentheses*

    – This operator tells the search engine to search first for what is grouped or nested inside the parentheses.
    – That is, ("kittens" OR "care") AND Siamese

Finally, the basic goals relate to navigation questions such as "Where am I?" Or "Where can I go?" (Applen 2002, p. 305). Moreover, such design approaches should involve user participation. Effective "communication and positive relationships must be cultivated and planned as any other successful component of project management" (Jiang et al. 2002, p. 20).

According to Issa (2008), navigation aims to determine the specific navigation paths through the website (including the new smart technology design) between the entities and to establish communication between the interface and navigation in the hypermedia application. Finally, navigation paths are "very important issues to address in website design, for the user has to be able to find what they are looking for as quickly as possible" (Darlington 2005, p. 75). The essential design techniques are: site, layout, link, and navigational structure for the hypermedia application.

5.4 Prototyping

Prototyping is considered a part of the development process and is used to evaluate different proposals for the final website or new smart technology design. Prototyping should be introduced in the new smart technology design (including devices, user interface, and website) to identify the layout and the potential problems in the early stages; "functional requirements; navigational issues and visual aspects can also be clarified with the aid of a prototype" (Darlington 2005).

Prototyping can be classified as evolutionary or throwaway. "Evolutionary, means that the prototyping becomes part of the final project," while throwaway prototyping "serves only as a pattern for implementation, and you can throw away the prototyping once the interface is complete" (McCracken and Wolfe 2004, p. 8).

Prototyping brings many advantages to the development process that improve communication in the system, including devices, user interface, and website, and to remove misunderstanding from requirements in order to demonstrate the object, action, or property being discussed, and to provide a basis for an ongoing debate with users about their system requirements. Finally, the prototyping approach place(s) greater emphasis on the interpersonal and communication skills of developers and users (Verner and Cerpa 1997).

There are two types of prototyping, namely: low fidelity and high fidelity. The latter will be similar to the final product of the website by using software such as Visual Basic, Smalltalk, and Macromedia, and it is recommended that more than one solution be produced (i.e., three solutions) in order to give the client more options about the "look" of the website. The advantages of high-fidelity prototyping are: It is very useful for detailed evaluation of the main design elements; it is useful for "selling ideas to people and for testing out technical issues" (Preece et al. 2002, p. 246), and it often constitutes a crucial stage in client acceptance—"as a kind of final design document which the client must agree to before the final implementation" (Benyon et al. 2005, p. 254).

Finally, low-fidelity prototyping does not look very much like the final product and uses materials that are very different from the intended final version; however, these prototypes are very useful since they tend to be simple, cheap, and quick to produce, i.e., storyboarding and sketching (Rudd et al. 1996).

Finally, Issa (2008) confirms that prototyping will allow users and management to interface with a prototype of the new website (including the new smart technology design) to gain some experience in using it. The aims of prototyping are to reduce cost and improve quality during the early stages in the development process.

5.5 Guidelines and Principles Design

To recognize the significance of HCI and usability features in the web development process as well as in the design process, it is worth scrutinizing the principles and guidelines of design suggested by Te'eni et al. (2007). The implementation of these principles and guidelines when designing and developing a new smart technology, device, user interface, including a website, will improve the presentation, performance, functionality, learnability, efficiency, effectiveness, usefulness, or utility; it will reduce errors and inaccuracies in the system, and this will lead to improved user satisfaction and achievement of the goals of both the designer and the user (Davis and Shipman 2011; Fernandez et al. 2011; Leung and Law 2012; Oztekin 2011).

To ensure that the design of a device, user interface, and website will match users' needs, design principles and guidelines are introduced and presented to designers. Principles are used to formalize the high-level and widely appropriate design goals while guidelines are essential to the designers to achieve the principles (Te'eni et al. 2007; Zhang et al. 2005). The design principles are divided into seven stages (see Fig. 5.1); each principle focuses mainly on a specific concept, which should be considered from the outset by the designers and users in order to develop a successful device or user interface including a website.

Fig. 5.1 Design principles (Adopted from Teeni et al. 2007—prepared by the Tomayess Issa)

The *design principles* are:

Improve users' task performance and reduce their effort: This principle aims to achieve high functionality along with high usability (i.e., efficiency, ease of use, and comfort in using the system, given that the functionality has been established). **Strive for fit between the information representation needed and presented**.

a. Representation: a simplified description of a real-world phenomenon.
b. Functionality: the set of activities.
c. Usability: a measure of ease of use.
d. Cognitive fit: system's representation of the problem supports the user's strategies for performing the task.

Direct and constraint user affordance to capture real-world knowledge: The general idea here is that the knowledge required to act effectively resides both in the person's head and in the real world around him/her.

Design for error: A faulty action due to incorrect intention (mistake) or to incorrect or accidental implementation of the intention (slip).

Designing for an enjoyable and satisfying interaction: The design of the interface or website should make the interaction enjoyable for both the designer and the users.

Promote trust is a critical component in developing an interface or website, especially for the e-commerce systems where the interactions translate directly into revenue.

Support diversity of users: This principle should take into consideration the diversity of populations of users.

To confirm that the device, user interface, or website is widespread and meets users' requirements, designers, especially HCI experts, must include these design principles in their agenda to prevent user frustration and dissatisfaction with these tools.

Furthermore, to ensure that the device, user interface, or website is well accepted by users, the designers and HCI experts must consider the design guidelines, which

Fig. 5.2 Design guidelines (adopted from Te'eni et al. 2007)—prepared by Tomayess Issa

are crucial in the web development process. The design guidelines comprise five steps (see Fig. 5.2).

The *design guidelines* are:

- *Consistency Guidelines*: If the interface is consistent (even if poorly designed), the end-user can adapt to it.
- *Control and feedback go hand in hand*: Providing feedback is probably the most accepted guideline in the design of any interaction.
- *Metaphor*: The use of familiar terms and associations to represent a new concept.
- *Direct Manipulation*: An interaction style in which objects are represented and manipulated in a manner analogous to the real world.
- *Design Aesthetic Interface*: Aesthetic appeal concerns the overall appearance of an application.

5.6 Evaluation and Testing

This section discusses the importance of the evaluation in the system development process for new smart technology, devices, interfaces, and websites. In general, evaluation is an essential step in the system development process, since experts and novices will evaluate the new smart technology, device, interface, or website and suggest solutions to problems (Jacobson et al. 1999; Nielsen and Molich 1990).

5.6.1 What is Evaluation?

Evaluation is intended to collect comments and evaluation from the users to ensure that devices, interfaces, and websites are meeting the users' needs (Issa 2008). To ensure that the functions of devices, interfaces, and websites are effective from the technical perspective, experts and novices test them using specific scenarios. According to McCracken and Wolfe (2004, p. 41), "expert- based evaluation can be achieved by using a group of usability experts to critique the prototype" while user-based evaluation can be performed by asking "users to perform representative tasks with the prototype."

Evaluation should occur in the initial stages of the system development process and prior to release to ensure that the device, interface, or website matches users' needs. Furthermore, evaluation takes place when the system is released and is used by target users in a real context, that is, during the use and impact stage.

In general, experts and users will evaluate new smart technology, devices, interfaces, and websites in terms of usability (i.e., efficient, effective, safe, utility, easy to learn, easy to remember, easy to use, easy to evaluate), HCI (usable, practical, visible, job satisfaction, additional features, text style, fonts, layout, graphics, and color), and navigation (site, layout, navigational structure for the hypermedia application) (Issa 2008).

5.6.2 Why Evaluate?

Additionally, designers, HCI experts, and users should understand the reasons for conducting evaluation. Preece et al. (1994) listed four main reasons as: (1) to understand the real world and how users employ the new smart technology in the workplace and social life, in order to provide further information to the designers to improve this new smart technology to better fit their needs and work and social environment; (2) to compare and contrast the new smart technology design in line to identify which is the best; (3) to determine whether the new smart technology design is matching the users, the projects goals, and the objectives; and finally 4) to check confirmation to a standard.

5.6.3 When to Evaluate?

In order to ensure that new smart technology design matches users' needs, the designers, HCI experts, and users should determine an appropriate time and means of conducting the evaluation. Currently, there are two approaches for formative and summative evaluation. Formative evaluation is conducted during the development of a product in order to form or influence design decisions. Summative evaluation is conducted after the product is finished to ensure that it possesses certain qualities, meets certain standards, or satisfies certain requirements set by the sponsors or other agencies (Hamilton and Chervany 1981; Nunamaker and Chen 1990; Shackel 1991).

5.6.4 Methods and Means of Evaluation

Real users in real-world contexts can conduct evaluation during the actual use of the produce, and this type is called "use and impact evaluation." However, the longitudinal evaluation aims to observe or examine a set of subjects over time with respect to one or more evaluation variables.

To have a successful evaluation, a plan should be formed to identify the stages of design (early, middle, late); the novelty of product (well-defined versus exploratory); number of expected users; criticality of the interface (e.g., life-critical medical system versus museum-exhibit support); costs of product and finances allocated for testing; and time available and the experience of the design and evaluation team (Gauthier 2015; Te'eni et al. 2007; Wakefield et al. 2015).

Examples of evaluation strategies include analytical methods (conducted by experts or designers to inspect potential new smart technology design problems) and heuristic evaluation (conducted by experts guided by a set of higher-level design principles or heuristics, evaluate to ensure if the new smart technology design is matching the principles and guidelines design). Furthermore, a guidelines review is conducted during the design stage with objective users (i.e., experts or designers outside the design team) to confirm whether the new smart technology design matches the project aims and objectives. Nielsen and Molich's ten (10) rules of thumb and Ben Shneiderman's eight (8) goals are considered the first step to establish the heuristics evaluation (Fard N.D.; Wong 2020) (see Fig. 5.3).

Additionally, cognitive walk-through evaluation is one of the evaluation strategies intended to identify the problems and glitches in the new smart technology design by asking the experts only to evaluate specific tasks in the design; on the other hand, the pluralistic walk-through evaluation will ask experts, designers, and users to examine the new smart technology design by considering specific scenarios. This type of evaluation is focused mainly on users' participation and how they would proceed with doing tasks.

In addition, in order to collect from users' further information about the new smart technology design, empirical methods are very useful used, i.e., survey/questionnaire, interviews, focus groups, laboratory experiments, and observing and monitoring usage through field studies. These methods are useful to obtain the necessary feedback from users to improve the new smart technology design and to match users' needs (Nielsen and Mack 1994; Shneiderman and Plaisant 2010).

Finally, according to Issa (2008), expert-based and user-based evaluations will test the website to ensure that the website functions effectively from the technical perspective. Functionality testing and evaluation are mainly about formative usability evaluation by experts and users.

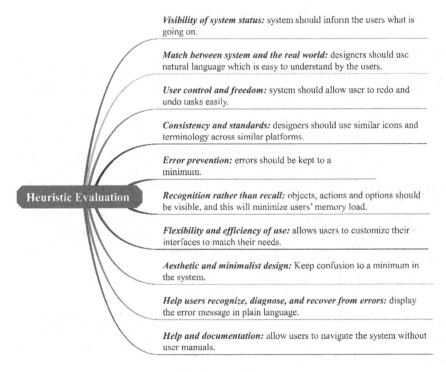

Visibility of system status: system should inform the users what is going on.

Match between system and the real world: designers should use natural language which is easy to understand by the users.

User control and freedom: system should allow user to redo and undo tasks easily.

Consistency and standards: designers should use similar icons and terminology across similar platforms.

Error prevention: errors should be kept to a minimum.

Recognition rather than recall: objects, actions and options should be visible, and this will minimize users' memory load.

Flexibility and efficiency of use: allows users to customize their interfaces to match their needs.

Aesthetic and minimalist design: Keep confusion to a minimum in the system.

Help users recognize, diagnose, and recover from errors: display the error message in plain language.

Help and documentation: allow users to navigate the system without user manuals.

Fig. 5.3 Heuristic evaluation, prepared by Tomayess Issa

5.7 Task Analysis

To develop a new smart technology which will help to make the devices very successful, the researchers need to incorporate additional detailed techniques. These will address specific deficiencies identified in the methodologies reviewed in the preceding sections. They relate to:

- Detailed task analysis (to facilitate a comprehensive set of links between the front end and back end of an e-commerce websites).
- Detailed procedures for website design and implementation.

It is very important to know one's users when an information system or a website is being developed. At the same time, the designer is required to acquire more information about what users will actually do. To answer this question, the designer needs to adopt a specific technique which is termed "task analysis." Task analysis is the "process of building a complete description of the [users'] (their) duties" (McCracken and Wolfe 2004, p. 44). This technique involves seeking the following information about the users:

- What tasks they perform.
- Why they perform them.

- How they perform them.

The information will assist designers to determine the basis and foundation for making decisions that will produce successful designs.

5.7.1 Goals, Tasks, and Actions

Participation by users is the basis for developing and creating a simple, easy-to-use user interface or website. Task analysis will help the designer to learn more about the goals and tasks of the users, and in turn to produce an interface that operates effectively and productively.

According to McCracken and Wolfe (2004), goals, tasks, and actions should be defined at the beginning of the project. Goals are work-related objectives that include searching for information, sending e-cards, registering a hotel guest, sending e-mail, or doing Internet marketing or non-work-related goals such as playing games, chatting, or making a plan. Therefore, goals "are technology independent, and they remain the same even when the technology changes" (McCracken and Wolfe 2004, p. 44).

On the other hand, tasks may or may not be consistent between users. Therefore, tasks need to be changed according to the users' requirements and needs, and these tasks are used to accomplish the goals (e.g., buying a book (about HCI) from Amazon.com).

Finally, the last step is action. Actions are "subcomponents of tasks" (McCracken et al. 2004, p. 44). In other words, actions are a series of steps which need to be followed in sequence in order to complete the tasks and, hence, achieve the users' goals. In addition, these steps may involve one or more sub-steps.

5.7.2 Techniques for Identifying Types and Granularity of Tasks

In this section, six techniques will be introduced which can be used to collect more information about the tasks, which are needed to achieve the users' goals. Sometimes, analysts may need to use more than one technique to collect information with respect to the tasks that are needed in order to accomplish the goals.

A key issue is "granularity." This refers to "the level of detail in a description" (McCracken and Wolfe 2004, p. 45). For example, users need to look at their tasks from a short distance to understand its detail as well as from a long distance, to know the purpose behind it. Therefore, in task analysis the granularity that is chosen will depend on "the nature and scope of your website development effort" (McCracken and Wolfe 2004, p. 45).

Workflow Analysis: The purpose behind this technique is to illustrate how the work will be done if more than one user is involved in the task. This means that this technique focuses "on work as it passes from person to person" (McCracken and Wolfe 2004, p. 46). As a result, this information may be vast and very helpful for the designer and user simultaneously as it provides a full picture of the project.

Job Analysis: This technique is the opposite of the former, as the designer needs to "focus on what a single person does in a day, a week, or a month" (McCracken and Wolfe 2004, p. 46). The designer can collect this information from the users by using the interview method or observing them in their work environment.

Task List: This technique takes "the granularity of job analysis to a more detailed level" (McCracken and Wolfe 2004, p. 46). In other words, the designer needs to think very carefully about how many tasks are to be studied in detail before these are broken down into more tasks. In addition, the designer should define and describe the components of a user's job, as some users are responsible for more than one job.

Task Sequences: This technique will establish "the order in which the tasks take place" (McCracken and Wolfe 2004, p. 47). The designer can learn the order of these tasks by observing the users at work. However, the important issue which needs to be taken into consideration is to try not to change the users' way of doing the tasks unless there is an important reason for doing so. It is better to give users full control to finalize their job in whatever sequence they like. However, "if you discover that a majority of users do things in a certain sequence, it makes sense to set up the interface to simplify things for the majority" (McCracken and Wolfe 2004, p. 47).

Task Hierarchies: The purpose of this technique is to document the components of a task, which are called sub-tasks. The level of detail depends on the type and the purpose of the website.

Procedural Analysis: This last technique "contains the most detail of any of the techniques" (McCracken and Wolfe 2004, p. 48). This step will give the designer information about how many steps need to be taken by the user in order to achieve his/her tasks.

Figure 5.4 shows that involving the users in this aspect of the system development process is essential in order to provide the necessary detailed information and to familiarize the users with the new system structure. However, the designer needs to take into consideration the level of user participation in the system development process, which means involving the users in one or more tasks during the process. The user participation level needs to be discussed by the designer and users so that an agreed process can be identified.

Finally, task analysis is essential in the new smart technology development process and involves determining the user types, their work goals, and activities, and applies to the device, user interface, and website.

Goals: are work-related objectives

Tasks: may or may not be consistent between users

Actions: are a series of steps which need to be followed in sequence in order to complete the tasks and hence achieve the users' goals

Granularity: refers to the level of detail in a description

Task Analysis

Fig. 5.4 Task analysis—prepared by Tomayess Issa

5.8 Conclusion

This chapter has discussed the issues of color, navigation, prototyping, principles and guidelines design, evaluation and testing, and task analysis in terms of new smart technology design. These design concepts are essential especially in new smart technology design, i.e., devices, user interface, and website. In general, color is widely used in the development process to attract users' attention and as reminders of specific information on a display. However, navigation enables the user to control the intersystem and intra-system flow of activities and the user's navigation of the system, while prototyping brings designs to life for both designers and users who will use the new design.

Furthermore, this chapter discussed the importance of the evaluation and testing of the new smart technology design as these aspects will assist users and designers to identify the problems and identify some solutions to prevent them in the future. On the other hand, to ensure that user interface or website is well accepted by designers and HCI experts must consider the design guidelines, which are crucial in the web development process. Finally, this chapter examined the task analysis focuses on goals, tasks, and actions of new smart technology design, and is concerned with logic, cognition, or purpose of tasks.

References

Adler JL, Blue VJ (1998) Toward the design of intelligent traveler information systems. Transp Res Part C Emerg Technol 6(3):157–172

Applen JD (2002) Tacit knowledge, knowledge management, and active user participation in website navigation. IEEE Trans Prof Commun 45(4):302

Benyon D, Turner P, Turner S (2005) Designing interactive systems: a comprehensive guide to HCI and interaction design, 2nd edn. Pearson Education Limited, UK

Blackmon MH, Polson PG, Kitajima M, Lewis C (2002) Cognitive walkthrough for the web. In: Proceedings of the SIGCHI conference on human factors in computing systems. ACM, pp 463–470

Cornforth D (1994) Color—its basis and importance. In: Quality attributes and their measurement in meat, poultry and fish products. Springer, pp 34–78

Cyr D, Head M, Larios H (2010) Colour appeal in website design within and across cultures: a multi-method evaluation. Int J Hum Comput Stud 68:1–21

Darlington K (2005) Effective website development : tools and techniques. Pearson Education Limited 2005, UK

Davis P, Shipman F (2011) Learning Usability Assessment Models for Web Sites. Paper presented at the IUI (2011) Palo Alto. California, USA

Elfes A (1987) Sonar-based real-world mapping and navigation. IEEE J Robot Autom 3(3):249–265

Elliot A, Maier MA (2012) Color-in-context theory. Adv Exp Soc Psychol 45:61–125

Fard A (N.D.) Heuristic Evaluation Checklist. https://uxlib.net/assets/attachment/Heuristic_Evaluation_Template_Adam_Fard_73611209ds.pdf. Accessed 2 Nov 2021

Fernandez A, Insfran E, Abrahão S (2011) Usability evaluation methods for the web: A systematic mapping study. Inf Softw Technol 53(8):789–817. https://doi.org/10.1016/j.infsof.2011.02.007

Fons J, Pelechano V, Albert M, Pastor O (2003) Development of web applications from web enhanced conceptual schemas. In: Conceptual Modeling-ER 2003. Springer, pp 232–245

Gauthier G (2015) A Usability Evaluation of a Website Focusing on the Three Initial Steps of the Conflict/Resolution Process for Union Members. http://scholarspace.manoa.hawaii.edu/bitstream/handle/10125/35855/Gauthier_Final_Paper_Scholarspace.pdf?sequence=1. Accessed 29 Sept 2021

Hamilton S, Chervany NL (1981) Evaluating information system effectiveness-Part I: comparing evaluation approaches. MIS Q 5(3):55–69

Holtze T (2006) The web designer's guide to color research internet reference service 11(1):87–101

Issa T (2008) Development and Evaluation of a Methodology for Developing Websites - PhD Thesis, Curtin University, Western Australia. http://espace.library.curtin.edu.au/R/MTS5B8S4X3B7SBAD5RHCGECEH2FLI5DB94FCFCEALV7UT55BFM-00465?func=results-jump-full&set_entry=000060&set_number=002569&base=GEN01-ERA02

Jacobson I, Booch G, Rumbaugh J, Rumbaugh J, Booch G (1999) The unified software development process, vol 1. Addison Wesley Reading

Jiang JJ, Chen E, Klein G (2002) The importance of building a foundation for user involvement in information system projects. Project Manag J 33(1):20

Kakumani BN, Brady EE, Klashinsky KB (2004) Method and system for using virtual labels in a software configuration management system. Google Patents

Labrecque LI, Milne GR (2012) Exciting red and competent blue: the importance of color in marketing. J Acad Mark Sci 40(5):711–727

Leung R, Law R (2012) Human factors in website usability measurement In: Zhang Y (ed) Future wireless networks and information systems, vol 143. Lecture Notes in Electrical Engineering. Springer Berlin Heidelberg, pp 501–507. https://doi.org/10.1007/978-3-642-27323-0_63

McCracken DD, Wolfe RJ (2004) User-centered website development: a human-computer interaction approach. Pearson Education Inc., New Jersey, USA

Morton J (2010) Why Color matters http://www.colorcom.com/research/why-color-matters. Accessed 27 Oct 2021

Morton JL (2015) Basic color theory color matters http://www.colormatters.com/color-and-design/basic-color-theory. Accessed 27 Oct 2021

Nielsen J, Mack RL (1994) Usability inspection methods. John Wiley and Sons NY

Nielsen J, Molich R Heuristic evaluation of user interfaces. In: Proceedings of the SIGCHI conference on Human factors in computing systems, 1990. ACM, pp 249–256

Nunamaker Jr JF, Chen M (1990) Systems development in information systems research. In: Proceedings of the twenty-third annual hawaii international conference on system sciences. IEEE, pp 631–640

Oztekin A (2011) A decision support system for usability evaluation of web-based information systems. Expert Syst Appl 38(3):2110–2118. https://doi.org/10.1016/j.eswa.2010.07.151

Pelet JE, Conway MC, Papadopoulou P, Limayem M (2013) Chromatic scales on our eyes: how user trust in a website can be altered by color via emotion. Adv Intell Syst Comput 205:111–121

Preece J, Rogers Y, Benyon D, Holland S, Carey T (1994) Human computer interaction. Addison-Wesley, England

Preece J, Rogers Y, Sharp H (2002) Interaction design: beyond human-computer interaction. John Wiley & Sons, USA

QSX Software Group (2015) Color Meaning. http://www.color-wheel-pro.com/color-meaning.html. Accessed 27 Oct 2021

Rudd J, Stern K, Isensee S (1996) Low vs. high-fidelity prototyping debate. Interactions 3(1):76–85

Shackel B (1991) Usability-context, framework, definition, design and evaluation. Human factors Inf ormatics usability:21–37

Shneiderman B, Plaisant C (2010) Designing the user interface: strategies for effective human-computer interaction. Addison Wesley, Boston, MA, USA

Sibagraphics (2015) The meaning of colours. http://www.sibagraphics.com/colour.php. Accessed 27 Oct 2021

Taylor RE, Sennott JW (1984) Navigation system and method. Google Patents

Te'eni D, Carey J, Zhang P (2007) Human computer interaction: developing effective organizational information systems. John Wiley & Sons Inc., USA

Verner JM, Cerpa N (1997) Prototyping: Does your view of its advantages depend on your job? J Syst Softw 36(1):3–16. https://doi.org/10.1016/0164-1212(95)00193-X

Wakefield B, Pham K, Scherubel M (2015) Usability evaluation of a Web-based symptom monitoring application for heart failure. West J Nurs Res 37(7):922–934

Wang L, Giesen J, McDonnell KT, Zolliker P, Mueller K (2008) Color design for illustrative visualization. Visual Comput Graph IEEE Trans 14(6):1739–1754

Wong E (2020) Heuristic evaluation: how to conduct a heuristic evaluation interaction design foundation. https://www.interaction-design.org/literature/article/heuristic-evaluation-how-to-conduct-a-heuristic-evaluation. Accessed 2 Nov 2021

Zhang P, Carey J, Te'eni D, Tremaine M (2005) Integrating human-computer interaction development into the systems development life cycle: a methodology. Commun Assoc Inf Syst 15:512–543

Chapter 6
Models and Methodologies

Abstract This chapter examines the various types of models and methodologies for developing systems (including websites), which may incorporate such HCI processes, usability, and Internet marketing issues. It assesses the advantages and disadvantages of each methodology and analyzes the differences between them in order to develop the framework for a new participative methodology. To produce a successful new smart technology, devices "system" (or website), both designers and users should be working collaboratively. Such user participation has to be facilitated by a system development methodology consisting of a clear sequence of stages and steps to be followed by the designer and participating users. The approach of breaking a methodology into stages and steps will be adopted in this research to facilitate the design process by breaking down the activities into several major stages and smallest parts into steps (within each stage).

6.1 Introduction

In order for new smart technology, devices, systems, (or websites) to be widely accepted and used effectively, they need to be well designed. To achieve this, designers and users need to use a specific methodology to produce the "system" (or website). A sound methodology is a very important component of the system development process, in order to produce a new system, which meets the user's requirements. A methodology "should tell us what steps to take, in what order and how to perform those steps but, most importantly, the reasons, 'why' those steps should be taken, in that particular order" (Jayaratna 1994).

The term "methodology" is used significantly in information systems development, as each methodology should have a set of stages and steps, which need to be followed in sequence if the work is to be done successfully. "Stage" is a "convenient breakdown of the totality of the information systems life cycle activity" (Olle et al. 1988, p. 21), while "step" is "the smallest part of a design process" (Olle et al. 1988, p. 21).

© Springer-Verlag London Ltd., part of Springer Nature 2022
T. Issa and P. Isaias, *Sustainable Design*,
https://doi.org/10.1007/978-1-4471-7513-1_6

The sequence of the stages may not always be fixed, but it "does suggest that there is a strict time scale applicable to all situations" (Olle et al. 1988, p. 30). In some projects, iteration between stages will occur and this may have a range of impacts on the methodology, as an iteration may "take different forms and thus impact differently on what one can do with a methodology" (Olle et al. 1988, p. 30).

The main demand is for methodologies that can lead to improvements in the following three aspects according to Avison and Fitzgerald (1993, p. 264): a better end product; a better development process; and a standardized process.

For that reason, a designer needs to understand users' requirements for the project before choosing the methodology, in turn to successfully complete the work and to accomplish profitable results.

In this chapter, Issa (2008) will discuss various types of models and methodologies, including: lifecycle models; IS development methodologies; methodologies with explicit human factors aspects; website methodologies; marketing methodologies; and additional techniques, such as task analysis[1] and detailed website design and implementation. There are numerous similarities in respect of the stages between methodologies for developing information systems, websites, or marketing strategies. Integrating stages from information systems methodologies into website and marketing methodologies is very beneficial in order to develop websites that are more effective and efficient. Human factors experts should be involved in these methodologies to make sure that transaction processes, tracking, maintenance, and updating of the website meet the users' requirements.

Firstly, Issa (2008) will discuss the methodologies in this sequence to identify two aspects: (1) the stages needed for the system development process and (2) the four key principles (user participation, usability, iteration, and real interaction), in order to check the availability of these four key principles in IS development, website, and marketing methodologies. The system's development cycle will be discussed in order to identify the stages.

Secondly, the stages of information systems development methodologies will be checked to assess how effectively they match the four key principles at each stage and to identify the strongest stage in each methodology. Thirdly, for the website and marketing methodologies, the researcher will: check the availability of techniques covering the four key principles in these methodologies; list the extra stages which will be added to the new methodology; and identify the strongest stage in each methodology.

Finally, additional techniques (i.e., task analysis[2] and detailed website design and implementation) will be discussed. The chapter will also identify any extra stages, which will be added to the new methodology, such as navigation, promotion, and staff training. Such additional detailed techniques will play a key role in the new methodology, as most of the existing methodologies have neglected these.

---

[1] Task Analysis: Please check Chap. 5.

[2] Task Analysis: Please check Chap. 5.

6.2 Lifecycle Models

The term "lifecycle model" is used to represent a model that captures a set of system development activities and how they are related (Preece et al. 2002). The more sophisticated lifecycle models inform the designer about when and how to move from one activity to the next and provide a description of the deliverables for each activity. These lifecycle models are popular since they allow developers, and particularly managers, to get an overall view of the development effort so that processes can be tracked, deliverables specified, resources allocated, targets set, and so on. As indicated, some lifecycle models include iteration—this "model incorporates iteration and encourages a user focus" (Preece et al. 2002, p. 186).

The stages in a typical development lifecycle model for interaction design are:

- Define the requirements.
- Prepare some alterative designs, which meet the needs, and requirements that have been identified previously.
- Select a preferred solution.
- Test and evaluate the design.
- Iterate, if necessary. This option can be used either before or after the evaluation stage.

This section discusses and compares a historical sequence of increasingly complex models (i.e., waterfall lifecycle model, spiral lifecycle model, and rapid applications development) which focus on interaction design and adopt the general approach of the development lifecycle model.

Furthermore, two models will be discussed in this section from the human–computer interaction perspective, the star lifecycle model, and usability engineering. The former focuses on how the designer addresses human–computer interaction design problems, while the latter "shows a more structured approach and hails from the usability engineering tradition" (Preece et al. 2002, p. 192).

6.2.1 The Waterfall Lifecycle Model

This model is basically a linear model where each stage must be completed before the next stage can be started. For example, requirements analysis has to be completed before design can begin. However, iteration can occur at each stage. This lifecycle model is divided into five sequential stages, which may be described as follows:

- **Requirements Analysis**: This stage begins when an organization seeks to add, improve, or correct a system, which is not meeting the requirements of the users. The requirements specification should be captured by the designer in consultation with users to know "what the eventual system will be expected to provide, and how the system will provide the expected services" (Dix et al. 1998, p. 181).

- **Design**: This stage will allow the designers to define the system specifications for the components, such as hardware and software, screen layouts, and documentation.
- **Code**: This stage involves converting design and system specifications into "executable programming language" (Dix et al. 1998, p. 182).
- **Test**: This stage will allow the users to test the new system to ensure that "the system meets their requirements" (Dix et al. 1998, p. 183).
- **Maintenance** :This stage involves the "correction of errors in the system which are discovered after release and the revision of the system services to satisfy requirements that were not realized during previous development" (Dix et al. 1998, p. 183).

One of the main flaws with this model is "that requirements change over time, as businesses and the environment in which they operate change rapidly"; hence, it "does not make sense to freeze requirements for months or years, while the design and implementation are completed" (Preece et al. 2002, p. 188). In addition, although a limited (between stages) iteration option is available in this model, the opportunity to constantly review and evaluate a proposed system with users is not included.

In practice, developing a website by using the waterfall model is complex since most of the users are not "clear how they would want the site to look" (Darlington 2005, p. 34). To solve this problem, prototyping should be introduced since it can help to identify the website layout and the potential problems in the early stages "functional requirements; navigational issues and visual aspects can also be clarified with the aid of a prototype" (Darlington 2005, p. 34).

6.2.2 The Spiral Lifecycle Model

For many years, the waterfall lifecycle model was considered the most popular model for the system development process. However, in 1988 Dr. Barry Boehm introduced the spiral lifecycle model. This model combines the waterfall model with an element called "risk analysis." It is divided into three major stages: (1) planning—to define the objectives, alternatives, and constraints; (2) risk analysis—for each of the alternatives solutions risks are identified and analyzed; and if this information is not enough, then the prototyping approach will be adopted, before; and finally, (3) engineering the solution.

This structured model is very useful as the customer can decide whether any one phase has been completed to his/her satisfaction before the next phase can commence. She/he may elect, if the risks are unacceptably high, to terminate the project. In addition, client evaluation can also be incorporated to check whether or not the system is developing according to their needs.

This model is very useful for large and complex development processes. The regular feedback from the customer allows for any necessary changes to be acted upon immediately. It incorporates steps to identify and control risks. This model "explicitly encourages alternatives to be considered, and steps in which problems or potential problems are encountered to be re-addressed" (Preece et al. 2002, p. 188). However, if not all aspects of risks are discovered in time, problems will surely occur, thereby leading to the need to repeat the procedures from the beginning and failure to meet the deadline for accomplishing the project. User involvement is not clearly defined in this model.

6.2.3 Rapid Application Development (RAD)

This model attempts to take a user-centered view and to minimize the risk caused by requirements changing during the course of the project by completing the stages as rapidly as possible. This model has five stages (namely project setup; JAD workshops; iterative design and build; engineer and test final prototype; and implementation review), and each must be completed before the next stage can be started. However, an iterative approach is incorporated, requiring the developer to go "back to the original data to gather and check the requirements" to determine whether or not it is supporting the user's tasks (Preece et al. 2002, p. 64). RAD added two new key features to the previous development models: time boxing and joint application development workshops.

- **Time boxing** breaks down a large project into many smaller projects. This will allow the designers to deliver the products incrementally and enhance flexibility in terms of the development techniques used and the maintainability of the final system.
- **Joint application development (JAD)** workshops between the users and developers are used to gain more information about any difficult issues that are faced and for decisions about system design to be made.

This model also specifically incorporates user testing of prototypes; however, it lacks maintenance of the implemented system. The prototyping in this model should be used to evaluate the system design and to identify the potential problems without any haste. Rapid development and manipulation of a prototype should not "be mistaken for rushed evaluation which might lead to erroneous results and invalidate the only advantage of using a prototype in the first place" (Dix et al. 1998, p. 207).

6.2.4 Systems Development Lifecycle

Kendall proposed the systems development lifecycle in 1992. This lifecycle is a "project management technique that divides complex projects into smaller, more easily managed segments or phases" (FFIEC IT Examination Handbook 2005). The segmentation of projects is a very useful method as it allows the designers and analyst to check if the previous stages have been successfully completed before moving to the next stage. This lifecycle is very constructive and useful as it prevents any tribulations to the designer, analysts, and users toward the end of the project.

This development lifecycle is divided into eight sequential stages (phases), with each needing to be completed before the next stage can be started. The stages are:

- **Initiation Phase:** This stage (phase) begins when an organization decides to add, improve, or correct a system, which is currently not meeting the requirements and needs for the organization and user simultaneously. Consequently, the management needs to define the following requirements before moving to later system development phases:

  - Business considerations (i.e., goals, objectives, budget, and legal issues).
  - Functional requirements (i.e., user requirements, hardware and software requirements and backup).
  - Project factors (i.e., project and risk management methodology, and estimated completion dates and costs).
  - Cost/benefit analysis (including both tangible and intangible benefits and costs)

(FFIEC IT Examination Handbook 2005).
All these requirements need to be considered and support documentation prepared before moving to the planning phase.

- **Planning Phase:** This stage (phase) is very significant as both designers and analysts need to study the requirements very carefully. Throughout this stage, the management needs to address the following items before shifting to the next phase: "communication, defined deliverables, control requirements, risk management, change management, standards, documentation, scheduling, budget, and testing and staff development" (FFIEC IT Examination Handbook 2005).
- **Design Phase:** This stage (phase) allows both the designers and analysts to carry out the design of the new system utilizing the requirements identified by the previous two phases. In this phase, initial prototyping is used to build mock-up designs of items such as applications screens, database layouts, and system architectures. This initial design needs to be reviewed by the users, designers, analysts, network administrators, and database managers to make sure it meets the requirements. The initial prototyping design is an iterative process, which means the system will remain in the stage and be reviewed by the participants "until they agree on an acceptable design" (FFIEC IT Examination Handbook 2005).
- **Development Phase:** This stage (phase) involves converting design specifications into an executable program (FFIEC IT Examination Handbook 2005).
- **Testing Phase:** This stage (phase) will allow the users to test the new system to ensure the accuracy of "programmed code, the inclusion of expected functionality and the interoperability of application and other network components" (FFIEC IT Examination Handbook 2005).

- **Implementation Phase:** This stage (phase) will involve installing the new system into the real-world environment. In addition, the users' training session for the new system will be carried out.
- **Project Evaluation:** This stage (phase) will allow the management to evaluate and review the "completion of the project objectives and assess project management activities" (FFIEC IT Examination Handbook 2005).
- **Maintenance Phase:** This stage (phase) involves changes and the correction of errors in the hardware, software, and documentation, which are discovered after the implementation stage.

According to Peters (1988), this lifecycle is a systematic breakdown of the software development process, "… A Software Life Cycle is both a management and a technical tool for organizing, planning, scheduling, and controlling the activities associated with a software development and maintenance effort" (cite in Jayaratna 1994, p. 33). However, this lifecycle does not allow for significant review and iteration between the stages; this means that suppleness and flexibility for responding to the particular needs of a specific project are missing. It also lacks detailed arrangements for user involvement at all stages.

6.2.5   The Star Lifecycle Model

The star lifecycle model was proposed by Hix and Hartson (1993) to address human–computer interaction issues in system development in a more flexible way. This model is six steps, namely implementation, task/functional analysis, prototyping, requirements specification, conceptual/formal design, and evaluation. This model incorporates two different modes of activity: the analytic mode and the synthetic mode. The former is described by concepts such as top-down, organizing, and working from the system view toward the user's view, while the latter is described by concepts such as bottom-up, free thinking, creative, and working from the user's view toward the systems view (Preece et al. 2002; Hix and Hartson 1993). The star lifecycle model is extremely flexible and popular, especially with managers, enabling them to get an overview of the "development effort so that process can be tracked, deliverables specified, resources allocated, targets sets and so on" (Preece et al 2002, p. 193).

The star lifecycle model can be adopted in any system development process, and the developer can move from any activity to any other without any specific order as the "activities are highly interconnected" (Preece et al. 2002, p. 193). The evaluation activity is at the center of this model, since, before moving to another activity, one need to pass through the evaluation activity to evaluate the result from the previous activity. This model can be used for defining requirements for a new system or for evaluating an existing situation and analyzing existing tasks. However, this lifecycle is very general and does not explicitly incorporate procedures for user participation or for system design and maintenance.

6.2.6 The Usability Engineering Lifecycle

Deborah Mayhew proposed the usability engineering lifecycle in 1999, and the purpose of this model is to focus more on how usability design and evaluation tasks may be performed alongside more traditional software engineering activities (Preece et al. 2002).

This lifecycle model presents a "menu of choices that can be worked into the broader development context in order to increase usability" (Instone 2004). It has three main aspects: requirements analysis, design/testing development, and installation. The production of a set of usability goals is the main aspect of the first stage since "these goals [are] captured in a style guide that is [then] used throughout the project to help ensure that the usability goals are adhered to" (Preece et al. 2002, p. 195). The middle stage in this model is the largest and most complex stage as many sub-tasks are involved to produce a detailed design. The final stage involves installation and user feedback.

The most important elements in the usability engineering model are experiential user testing and prototyping, combined with iterative design. "Because it's nearly impossible to design a user interface right the first time, we need to test prototype and plan for modification by using iterative design" (Nielsen 1992, p. 13).

It is anticipated that, via this lifecycle, the software engineering discipline "will embrace and incorporate usability engineering and it will become widely institutionalized in development organizations, similarly to how software engineering methodologies in general have become institutionalized" (Mayhew 1999, p. 33). However, this explicitly "human factors" approach is not easily integrated into the more general technical aspects of other models. This needs to be accomplished by operationalizing the model by using a methodology.

6.2.7 Summary of Lifecycle Models

Several stages were discussed in the lifecycle models section. The stages that are essential for the development of an information system interface, or website, can be summarized as planning, analysis, design, testing, implementation, evaluation, and maintenance. These stages are vital if the designer is to develop an interface, new smart technology, or website, which meets the user requirements and needs. However, the models need to be operationalized as detailed methodologies. As discussed in Chaps. 2 and 3, a critical aspect of systems development is effective HCI; hence, methodologies must adequately address this aspect. Four key principles (user participation, usability, iteration, and real interaction) are identified as fundamental aspects in order to develop systems in an effective manner by involving users from the beginning. The four key principles are considered the main foundation for this research to produce websites with high usability, thereby:

- Involving the users in the design from the beginning.
- Avoiding frustrations for the users.
- Making the website more approachable, friendly, and interesting.
- Winning the trust of the site visitors by meeting users' requirements.

The four key principles are:

- *User participation:* The main purpose is to allow user participation in the website development process to gain more information about the problems, elicit alterative solutions from the users, and familiarize them with the website before it is released.
- *Usability:* To confirm that the website design is efficient, effective, safe, has utility, easy to learn and easy to remember, usable, practical, provides job satisfaction, and incorporates performance measures that effectively assess the users' requirements and requests.
- *Iteration:* To allow for effectiveness and self-correction, this approach will assist the designers to build up the new website and ensure that the project will be tested repeatedly until it meets users' requirements.
- *Real Interaction:* The designer will track users' behavior to present statistics and useful information to demonstrate what attracts or repels users. This can be achieved by adding two options to the web: (1) feedback form to outline users' needs or (2) adding a counter to a webpage, which will provide detailed statistics (log file) to the designer. Information obtained will include which "Web pages are viewed most often, which domains request web pages, and what paths users follow as they navigate through a site" (Lazar 2006, p. 44).

In the subsequent sections, the presence of these aspects will be reviewed for each methodology. The rating used for these four key principles will be from 0 to 3. The former presents zero availability, while the latter is the maximum. Ratings of 1 and 2 indicate that these aspects are covered in a minimal or moderate way, respectively.

6.3 Information Systems Development Methodologies

System development lifecycle models may be operationalized using methodologies. Information systems development methodologies (ISDMs) are an "organized collection of concepts, methods (or techniques), beliefs, values, and normative principles supported by materials resources" (Iivari et al. 2001, p. 186). The main purpose behind using an ISDM is to guide the designer in performing the work by following specific stages in sequence. When developing a system or website, the analyst needs to study the different types of methodologies in respect of their similarities and differences and select the methodology, which best meets the project requirements.

Avison and Fitzgerald (1993) describe the status of information systems development methodologies as a "methodology jungle." This status of ISDM is "an unorganized collection of numerous methodologies which are more or less similar to each other" (Hirschheim et al. 1998). It was estimated that more than "1000 brand-named methodologies are in use all over the world" (Jayaratna 1994, p. xvii).

It is very difficult for the designer to review the vast array of existing ISDM and check which methodology will accomplish the work to be done. Therefore, the most important aspect of developing a new methodology is "to understand the existing stock and the collective methodology knowledge embedded in them" (Hirschheim et al. 1998). A new methodology should not merely duplicate an existing one but should offer some positive improvement. Consequently, this researcher will develop a new participative methodology for developing websites from the marketing perspective by embedding and grafting stages from various methodologies (Jayaratna 1994) such as those for developing information systems, websites, and marketing plans.

Various types of methodologies will be discussed in this section from perspectives of the information systems, human–computer interaction, and websites: Structured Systems Analysis and Design Methodology (SSADM); Soft Systems Methodology (SSM); User-Centered Development Methodology; and ETHICS. These methodologies have been chosen for assessment as they cover a range of perspectives, which are likely to address the four key principles identified above.

Such methodologies lay out specific stages to be undertaken and incorporate a range of principles from the lifecycle models discussed in the previous section. This will be presented in a table at the end of each methodology section to address two aspects: (1) checking the level of availability of techniques covering the four key principles in each stage of the methodology and (2) identifying the strongest stage for each methodology. This information will help the researcher in two aspects: (1) to recognize the importance of these four key principles in particular methodologies and (2) to select stages that will promote the structure of the new participative methodology for developing websites.

6.3.1 Structured Systems Analysis and Design Methodology (SSADM)

This methodology gives the designer "very detailed rules and guidelines to work to" (Avison and Fitzgerald 1993, p. 191), and "techniques, documentation and training procedures for developing information systems" (Avison and Wood-Harper 1990, p. 181). This methodology is classified into two major parts: three stages of systems analysis and three stages of systems design. The purpose behind this classification is to "make it easier to judge the proportion of time to spend on analysis" (Avison and Fitzgerald (1993, p. 192). Thus, this methodology is divided into six sequential stages, each of which needs to be completed before the next can be started. The stages are as follows:

- **Analysis of the current system** investigates and defines the problems of the current system.
- **Specification of the required system** defines the aims and services of the new system.
- **User selection of service levels, including technical options:** This stage focuses on users' participation and a feasibility study.
- **Detailed data design:** To define data and the relationships between them and to ensure that the data model meets the requirements of the individual users and the client organization.
- **Detailed procedure design:** This stage is the trial design for the system. The prototype can be paper based. The user will check if the trial design is working according to their requirements.
- **Physical design control** develops the system from the paper prototype to an implemented system. The users can use it and test the final system.

One of the main flaws of this methodology is that it cannot adequately "address the problem of project control and estimating costs directly through the incorporation of project management tools" (Avison and Fitzgerald (1993, pp.202–203). In addition, there is limited provision for iteration between stages and maintenance is missing. Table 6.1 indicates that user participation is moderate in the analysis stage. There is only a minimum rating for user participation and iteration aspects in the design stage to ensure that the data outcomes meet user requirements. Usability and real interaction aspects are rated as zero for each stage of this methodology. The strongest stage in the SSADM methodology is the design stage. This stage will help to identify the data and the relationships between them and produce the trial design for the system. The trial design will be checked by the users to assess if it is working according to users' requirements and requests.

Table 6.1 Structured systems analysis and design methodology—prepared by Tomayess Issa

Stages Principles	Planning	Analysis	Design	Testing	Implementation	Evaluation	Maintenance
User participation	0	2	1	2	0	0	0
Usability	0	0	0	0	0	0	0
Iteration	0	0	1	0	0	0	0
Real interaction	0	0	0	0	0	0	0
Strongest stage in SSADM	–	–	☑	–	–	–	–

6.3.2 Soft Systems Methodology (SSM)

Checkland proposed the Soft Systems Methodology (SSM) in 1981. SSM provides a "way of tackling messy situations in the real world" (Checkland and Scholes 2003, p. 1). A powerful argument in favor of SSM is that it "has been found to be transferable to people beyond those who developed it, and has been used in several hundred projects around the world" (Rosenhead and Mingers 2002, p. 112). According to Checkland, the SSM involves three roles: client, problem solver, and problem owner. The "client" "is the person or persons who caused the study to take place" (Checkland and Scholes 1990, p. 47), while the "problem solver" "wishes to do something about the situation in question, and the intervention had better be defined in terms of their perceptions, knowledge and readiness to make resources available" (Checkland and Scholes 1990 p. 47). The "problem solver" is responsible for turning the proposals for change "into real-world action in doing the study" (Checkland and Scholes. 1990 p. 48). The "problem owner" is the person/group for whom the system has consequences. This methodology is divided into seven sequential stages where each stage must be completed before the next stage can be started.

The stages are as follows:

- **Problem Situation Unstructured (1)**: The purpose of this stage is to define the problem and to gain more information and understanding of the problem in general; for example, the SSM should recognize the organization's culture and policies. This can be achieved by meeting the members of the organization and gaining as much information as possible about the organizational structure and culture.
- **Problem Situation Structured (2)**: At this stage, the analyst evaluates the problem situation from various approaches and by different stakeholders, in other words, to examine and assess the situation from different worldviews. The stage has several steps: intervention analysis, social and cultural analysis, political analysis, rich picture, and utilizing formal and informal methods.
- **Intervention Analysis:** This step will help the analyst to define the three roles through which they will learn more about problem situation in general:
  - *Client* "is the person or persons who caused the study to take place" (Checkland and Scholes. 2003, p. 47).
  - Problem solver solver defines the problem solver, resources, and the constraints.
  - *Problem owner:* No one is intrinsically a problem owner. The problem solver must decide who is to take [the role of] possible "problem owner" (Checkland and Scholes. 2003 p. 47). In addition, the problem owner is the entity "who has a feeling of unease about a situation, either a sense of mismatch between 'what is' and 'what might be' or a vague feeling that things could be better and who wishes something were done about it" (Checkland 1981, p. 294).

- **Social** and **Cultural Analysis:** This step will help the analyst to know more about the internal policies of the organization and to learn more about the motivation and features that effect an individual at the organization. Under this stage, the analyst needs to think about relevant roles, norms, and values, as these behaviors nor are fixed, they changed "steadily through time, sometimes slowly sometimes remarkably quickly" (Checkland 1981, p. 231) according to the situation:
- **Roles**: "a social position recognized as significant by people in the problem situation" (Checkland and Scholes 2003, p. 49)

  - *Norms* is a "specific prescriptions and proscriptions of standardized practice" (Checkland 1981, p. 231).
  - *Values* is an "express preferences, priorities or desirable states of affairs" (Checkland 1981, p. 231).

- **Rich picture**: is a graphical representation and communication model between the analysts and users to understand system problems and how they can be solved.
- **Formal and Informal Methods:** This step will help the analyst to collect more information about the system by using various methods, informal and formal, such as work observation, interviews, workshops, and discussions.
- **Naming of Relevant Systems (3):** This stage aims to involve system-thinking activities. In other words, this stage will involve "formulating of root definitions to a number of relevant systems" (Checkland and Scholes 2003, p. 33). This stage has several steps, such as root definition and CATWOE analysis, which are very important steps as they focus on the human activity systems.

  - **Root Definition:** Checkland and Scholes (2003, p. 33) define root definition as a way to "expresses the core purpose of purposeful activity system." In other words, the core purpose is the transformation process in which some entity "the input" changes into a new form of entity "the output." There are two kinds of root definition supported in SSM: "primary task root definition" and "issue-based root definition." The latter is concerned with one-off occurrences (such as a management restructuring), while the former is part of regular activities in the organization.
  - **CATWOE analysis** is a way to provide the analyst about with the structure of the real-world situation by answering "six element who is doing what for whom, and to whom are they answerable, what assumptions are being made, and in what environment is it happening?" (Avison and Fitzgerald (1993, p. 247). In other words, "a root definition meeting CATWOE requirements would have driven us more quickly towards aspects which with hindsight we know were finally crucial; we got there in the end, but with CATWOE we should have been quicker" (Checkland 1981, p. 226). According to Checkland and Scholes (2003 p. 35), CATWOE stands for:

    C: "Customers": the victims or beneficiaries of system activities.
    A: "Actors": people who do the activities.
    T: "Transformation": the conversion of input to output.

W: "Weltanschauung": the world view which makes this definition meaningful.

O: "Owners": those who can close the system or stop the event from happening.

E: "Environment": elements outside the system, which it takes as given.

Two of the major things, which need to be considered, are the transformation (T) and weltanschauung (W). The analyst needs to take care with respect to the transformation (T) as it is "frequently misunderstood, and the systems literature is full of inadequate representations of system inputs and outputs" (Rosenhead and Mingers 2002, p. 74). Moreover, the weltanschauung (W) might be extreme, such as a "terrorist system" or "freedom-fighting system" (Checkland 1988 p. 244). Therefore, it is essential to declare a "world view when giving an account of any purposeful activity" (Checkland 1988 p. 244).

- **Building the Conceptual Model (4):** This stage is unique and important as it is considered the core of the SSM. It is now required to establish the system requirements from the information, which was gathered from the previous stages. The conceptual model is used as "debating point so that the actors can relate the model to the real-world situation. Usually there is a conceptual model drawn for each root definition and the drawing up of several root definitions and conceptual models becomes an iterative process of debate and modification towards an agreed root definition and conceptual model" (Avison and Fitzgerald (1993, p. 247). The stage has several steps: formal system thinking and monitoring the system.

  - *Formal system thinking* serves as a guideline for checking the conceptual model to determine whether or not it meets the user's requirements.
  - *Monitoring the System:* This step will assist the analyst to monitor the system by defining three activities: (1) evaluating the performance in respect of efficacy, efficiency, and effectiveness; (2) monitoring the activities in relation to the problem definition; and (3) taking control action.

- **Comparison (5):** In this stage, the analyst will compare the conceptual models developed in stage four (4) with the definition of the problem situation in stage two (2). The purpose behind this comparison is to define and analyze the differences and similarities between the model and the real world in order to have a "well-structured and coherent debate about a problematical situation in order to decide how to improve it" (Checkland and Scholes 2003, p. 42).

- **Definition of Desirable and Feasible Changes (6):** This stage is important as the analyst will define those changes that are most feasible and desirable, bearing in mind such considerations as cost and benefit behind the change. It is very important to take into consideration these issues especially before the implementation stage in order to have positive outcomes, which meet the system needs.

- **Recommended Action (7):** This stage defines the changes to the system, and these recommendations should have the approval of the top level in the management before the implementation.

This methodology is a flexible process, as most of the stages can be iterated within the process if improvement is needed. The Soft Systems Methodology seeks to "create a system of enquiry which may be used to examine problem situations and lead to action decisions at both the level of what is required, and how the requirement can be met" (Cropley and Cook 1999, p. 4).

The SSM was created to support the human factors activities in complex existing and new systems. SSM is useful for two reasons: (1) Tt "bring clarity to confused situation and finding systems solutions in the world of human affairs using "systems"" (Checkland 2000, pp. 807–813); (2) it helps an organization to allow their systems "less fragmented, less random, more organized, more capable of generating insights and producing commitments" (Checkland 2000, p. 823). This methodology is not appropriate for all situations, as it requires a large gathering of information and often it involves human factors in various stages of the methodology. This methodology is useful when the objectives for the new system need to be clearly defined and clarified, and perhaps the most important issue is how the objectives can be accomplished, via a high-level approach. However, this methodology does not provide for the development of detailed specifications or testing of the system, especially regarding technical aspects. It handles organizational human factors well but does not address detailed design or evaluation of user interfaces.

Table 6.2 indicates that user participation is moderately well utilized in the early stages. Iteration is available in all of the stages with minimum availability to assess if improvement within the system is needed. In contrast, there is a zero rating for usability and real interaction in this methodology. The strongest stages in SSM are planning, analysis, and design. The planning stage examines the nature of the requirements for change and assesses how to address them. The analysis stage will require the analyst to perform the following: (1) evaluate the problem from different angles and from the view of different stakeholders; (2) evaluate the internal policies of the organization; (3) present a graphical presentation (called "rich picture") to the current situation to understand the problem in the system and how to be solved it; and (4) more informal and formal tools will be used to collect information about the

Table 6.2 Soft systems methodology (SSM)—prepared by Tomayess Issa

Stages Principles	Planning	Analysis	Design	Testing	Implementation	Evaluation	Maintenance
User participation	1	3	2	0	0	0	0
Usability	0	0	0	0	0	0	0
Iteration	1	1	1	1	1	1	1
Real interaction	0	0	0	0	0	0	0
Strongest stage in SSM	☑	☑	☑	–	–	–	–

system through observation, interviews, workshops, and/or discussion, while in the design stage, a small number of considerations should be addressed to identify the purpose behind establishing this system such as: (1) what the system is; (2) how the system will work; and (3) the purpose behind using this system. In addition, users will be involved in the system design and participate in the decision making.

6.3.3 User-Centered Development Methodology

Another methodology, which may be used to develop successful user interfaces for information systems, is the User-Centered Development Methodology. From the denotation, we learn that this method focuses on involving the user in the process as much as possible, with the ambition that the interface should meet the user's expectation. This can be achieved by user participation within the process activities, such as "observing users while they work, inviting users to participate on the design team and asking users to try out the product and following up on their feedback" (McCracken and Wolfe 2004, p. 5). This methodology involves numerous stages, which focus on "gathering information, designing, building, and testing of a prototype of the interface" (McCracken and Wolfe 2004, p. 5). It is divided into eight sequential stages, with each needing to be completed before the next stage can be started. The stages may be described as follows:

- **Needs Analysis**: Defining the purpose of developing the interface (or website).
- **User and Task Analysis**: Defining the users' type and the type of work users will do with the user interface or the website. User and task analysis focuses on user's goals and their activities, which are carried out by them to achieve their goals. For example, user analysis needs to define age, education level, and user computer knowledge. Task analysis examines user goals. McCracken and Wolfe (2004, p. 7) state that "many products fail because the development team didn't take the time to find out who their users are or what they want to do."
- **Functional Analysis**: Defining the functions, which will be available in the interface. Through these functions, the users will define their activities in order to achieve their goals.
- **Requirements Analysis**: Defining the "formal specifications (i.e., Data Dictionaries, Entity-Relationship Diagrams, and Object-Oriented Modeling) required to implement any system, including websites" (McCracken and Wolfe 2004, p. 7).
- **Setting Usability Specifications**: Defining what usability means for the interface. For example, "performance measure" (i.e., "number of tasks completed," "number of errors," "first impression," and "overall Satisfaction") (McCracken and Wolfe. 2004, p. 7).
- **Design**: Defining the appearance of the interface, which means, defining the content of the interface and to "organize it according to your user's exceptions."

The design "includes the layout of individual pages and how to use visual organization techniques to create clarity and consistency between pages" (McCracken and Wolfe 2004, p. 7).

- **Prototyping**: Developing the initial version of the interface. Prototyping can be classified as evolutionary or throwaway. "Evolutionary, means that the prototyping becomes part of the final project," while throwaway prototyping "serves only as a pattern for implementation, and you can throw away the prototyping once the interface is complete" (McCracken and Wolfe 2004, p. 8).
- **Evaluation**: Testing the interface by using expert-based evaluation and/or user-based evaluation. According to McCracken, "expert- based evaluation can be achieved by using a group of usability experts to critique the prototype" while user-based evaluation can be performed by asking "users to perform representative tasks with the prototype" (McCracken and Wolfe 2004, p. 8). Formative evaluation means "evaluation done during design to check that the product continues to meet users' needs" (Preece et al 2002, p. 323).

This methodology is "highly iterative and involves as much testing and revision as possible" (McCracken and Wolfe 2004, p. 5). This cycle of repetition can occur in the design, prototype, and evaluation steps, and will be successively run until the interface meets the usability specifications. The most important step is to take into consideration user goals and their tasks, as by missing this step, the results will be unsuccessful and unproductive. On the other hand, two basic concepts are missing in this methodology—that is, implementation and maintenance stages. It is also focused on the detail of user interface design without examining the overall relationship between social and technical aspects of the proposed system.

Table 6.3 demonstrates that the four key principles are available in numerous stages with ratings raging from minimum to maximum. User participation is incorporated in analysis, design, testing, and evaluation stages. Testing and evaluation stages are important to ensure that the system meets user requirements. Iteration has minimum rating in design; testing; and evaluation stages. Usability aspects are well

Table 6.3 User-centered development methodology (UCDM)—prepared by Tomayess Issa

Stages / Principles	Planning	Analysis	Design	Testing	Implementation	Evaluation	Maintenance
User participation	0	1	1	1	0	2	0
Usability	0	0	3	3	0	3	0
Iteration	0	0	1	1	0	1	0
Real interaction	0	0	0	0	0	0	0
Strongest stage in UCDM	–	☑	☑	☑	–	☑	–

covered to ensure user satisfaction with the interface. Finally, the real interaction has zero rating in this methodology.

The strongest stages in the User-Centered Development Methodology are analysis, design, testing, and evaluation. The analysis stage will help the analyst to identify the user's type, goals, and the activities, which are carried out by them to achieve their goal. The design stage will define the appearance of the interface. Testing and evaluation stages are included in this methodology, as the interface will be tested by expert-based and user-based evaluation to ensure that the interface or website meets user's requirements.

6.3.4 ETHICS Methodology

Mumford defines a specific methodology with high levels of stakeholder participation called "ETHICS," standing for "Effective Technical and Human Implementation of Computer-based Systems" (Mumford 1995, p. 3). Designers need to involve the user from the beginning, to keep focused on the target audience, to evaluate their activities, and to see if they "address the needs of the contemporary consumer" (Boyer 1999, p. 113). Users, through involvement in the development process, may be able to help to "shape design decisions in ways that deal with their concerns or make their work easier" (Doll and Torkzadeh 1989, p. 1156).

Participation is central to the ETHICS methodology as Mumford defined it as "handing responsibility for the design of a new system to the employees who eventually will have to operate it" (cited in Flynn 1992, p. 300). Two arguments were established from this definition. The first argument is user participation, which needs to be a part of the system development process, whether it be a new or existing system, so that decisions can be made which concern the purpose of the new system. User involvement in the design task can be through groups: "Involvement requires the creation of participative groups, and decisions on the amount and nature of their contribution to the total design process must therefore be made" (Mumford 1995, p. 50).

The second argument is the socio-technical approach that is mainly focused on increasing the ability of the individual to "participate in decision making and, in this way, to enable him/her to exercise a degree of control over the immediate work environment" (Mumford 1996, p. 70). The members of the Tavistock Institute for two specific reasons created this approach: to create "democratic organizations that are excellent in both human and production terms" (Mumford 1996, p. 73) and to consider the interaction between the social and technical parts of any work system. User involvement in the system development process, according to Mumford, "produces productivity, quality, coordination and control; but also provides a work environment and task structure in which people can achieve personal development and satisfaction" (cited in Flynn 1992, p. 301). Designing and implementing the social–technical approach is not an easy task, as it requires involvement from the users and management simultaneously. Furthermore, this approach requires "training, information, good administration, and skill" (Mumford 1996, p. 77). By adopting these approaches in the new system development process, the outcomes will offer benefits in respect of users' job satisfaction and success of an enterprise.

ETHICS is "pragmatically oriented and relies for its success on the practical abilities and the commitment of the participants to arrive at consensus decisions. It aims to build computer-based information systems which provide job satisfaction and met the efficiency needs of the organization" (Jayaratna 1994, p. 152).

The ETHICS methodology has three objectives focusing on the management of change. These objectives concentrate on the users and their participation in the computer system.

Firstly, the users play a major role in the design of the system, to enrich both job satisfaction and efficiency gains. Mumford said user groups with job satisfaction are able to cope with the required job changes and are "better able to diagnose their own job satisfaction needs than any outside group of specialists" (Mumford 1995, p. 3). An efficiency gain concentrates on user knowledge and the experience in dealing with these interfaces. This experience can be gained by dealing with these interfaces daily, learning about the user needs and system problems. Therefore, users can make a "useful contribution to the specification of the former and the solution of the latter" (Mumford 1995, p. 3).

Secondly, the users are encouraged to contribute to the system design, to define and set satisfaction objectives, and to supply additional information to the designer to aid in solving the problems within the system. In addition, the user can contribute his/her experience to explanations of "usual technical and operational objectives" (Mumford 1995, p. 3).

Thirdly, the ETHICS methodology can help ensure that the new technical system is surrounded "by a compatible, well-functioning organizational system" (Mumford 1995, p. 3). This objective is covered by the following concepts:

- Design of work procedures and instructions, for individual work or within groups.
- Define the relationship between the departments or functional areas which the new system will affect.
- The creation of good boundary management techniques.
- Focus on internal and external customers' needs.

(Mumford 1995, p. 4)

The ETHICS methodology is basically a linear model where each stage must be completed before the next stage can be started. It involves definition of a set of system characteristics including why change is needed; systems boundaries; description of the existing system; definition of the key objects and tasks; key information needs; diagnosis of efficiency needs; diagnosis of job satisfaction needs; design of the new system; technical options; preparation of detailed design work; and implementation and evaluation (Jayaratna 1994).

This methodology recommends many guidelines which are useful for "the understanding and the design of human-centered systems" (Jayaratna 1994, p. 174), and

Table 6.4 ETHICS methodology—prepared by Tomayess Issa

Stages Principles	Planning	Analysis	Design	Testing	Implementation	Evaluation	Maintenance
User participation	2	3	3	3	3	3	3
Usability	0	0	0	0	0	0	0
Iteration	0	0	0	0	0	0	0
Real interaction	0	0	0	0	0	0	0
Strongest stage in ETHICS	–	☑	–	–	–	–	–

to achieve improvements in efficiency, effectiveness, and job satisfaction in the work environment. ETHICS is a "participative design strategy and so employees and users will always be involved in analyzing needs and problem and deciding on solutions" (Mumford 1995, p. 78).

However, the main flaws of this methodology are its inability to handle the "interpersonal and political conflicts that may arise from opening up human feelings and emotions" and its lack of any means, "of discussing or resolving many of the ethical dilemmas that could arise in system development" (Jayaratna 1994, p. 174). In addition, it is quite hard for unskilled users to do the design work appropriately when using this methodology. This methodology does not incorporate iteration between stages, for detailed technical analysis and design or for maintenance.

User participation is dominant in this methodology, to enrich both job satisfaction and efficiency gains. However, there are zero ratings for usability, iteration, and real interaction in this methodology. The strongest stage in the ETHICS methodology is the analysis stage. This stage defines the user needs and problems, which allow the analyst to develop a system, which meets the users' requirements and their objectives (see Table 6.4).

6.3.5 Summary of Information Systems Development Methodologies

This section will provide a summary behind the information systems development methodologies.

For example, in the **Structured Systems Analysis and Design Methodology (SSADM)** only user participation and iteration stages are available in the design stage, while there is a zero rating for usability and real interaction. The strongest stage in SSADM methodology is the design stage, as this stage will help to define the data and the relationships between them and produce the trial design for the system.

In the **Soft Systems Methodology (SSM),** numerous techniques for user participation and iteration are available, while there is a zero rating for usability and real interaction. The strongest stages in the SSM are analysis and design. The purpose behind the analysis stage is to evaluate the situation from different angles and to collect more information to understand the system problem, so as to solve it. The main focus of the design stage is to determine the purpose of establishing this system and involving the user in system design and decision making.

User-Centered Development Methodology is different from the above methodologies as the four key principles are available in various stages with different ratings, ranging from minimum or maximum availabilities. The most dominant key principle in this methodology is usability to ensure that the interface is easy to learn, easy to use, and with less error frequency, while the real interaction has zero rating in this methodology. The strongest stages in the User-Centered Development Methodology are analysis, design, testing, and evaluation. The analysis will define the type of user(s) and their goals and activities, while the design stage will define the development of the interface. Experts and users combine testing and evaluation phases in one stage to test the interface. Finally, with the **ETHICS methodology,** only the user participation aspect is available, to enhance both job satisfaction and efficiency gains, while zero ratings are given for usability, iteration, and real interaction. The strongest stage in the ETHICS methodology is analysis, as via this stage, the analyst will define the users' needs so as to allow the new system to meet their requirements. Table 6.5 illustrates the strongest stages from the information systems development methodologies analyzed in this chapter and lists the rating availability for the four key principles in each stage. After reviewing the information systems development methodologies and studying each stage, it was noticed that implementation and maintenance were not considered the strongest stages for any of these methodologies, since the focuses of these methodologies are:

- Defining the system problem and clarifying users' needs for the new system.
- Evaluating the current situation and collecting more information to solve the system problem.
- Defining the relationships between the information and produce the trial designs for the system.
- Testing and evaluating the system to ensure that it meets the users' requirements.

However, techniques for effective implementation and maintenance of information systems are included in other (more technical) information system development methodologies not considered above. Since the objective is to develop a methodology for websites, it will be more effective to seek implementation and maintenance techniques targeted to websites. This is addressed in the next section.

Table 6.5 Summary of strongest stages in information systems development methodologies—prepared by Tomayess Issa

Stage	Information systems development methodologies	Principles			
		User participation	Usability	Iteration	Real interaction
Planning	Soft Systems Methodology	1	0	2	0
Analysis	Soft Systems Methodology (SSM)	3	0	2	0
	User-Centered Development Methodology (UCDM)	1	0	0	0
	THICS methodology	3	0	0	0
Design	Structured Systems Analysis and Design Methodology (SSADM)	1	0	1	0
	Soft Systems Methodology (SSM)	2	0	2	0
	User-Centered Development Methodology (UCDM)	1	1	3	0
Testing	User-Centered Development Methodology (UCDM)	1	1	3	0
Implementation	-	–	–	–	–
Evaluation	User-Centered Development Methodology (UCDM)	2	1	3	0
Maintenance	–	–	–	–	–

6.4 Methodologies for Developing Websites

There are many similarities between methodologies for developing information systems and websites. However, there are also differences. In this section, a range of methodologies from the website's perspective will be discussed in detail, including Human Factor Methodology for Designing Websites; Relationship Management

Methodology (RMM); W3DT Design Methodology; Information Development Methodology for the Web; and the Web Site Design Method (WSDM). This discussion will define the stages, which need to be carried out, by the designer and users in order to design a website, which meets the user requirements. Most stages focus on feasibility, navigation, deployment, promotion, and measurement of usability and effectiveness.

At the end of each methodology, the researcher will present a table showing: 1) the ratings for the four key principles in each stage within the methodology; 2) the strongest stage for each methodology for developing websites; and 3) the extra stages available in each methodology. These extra stages will add effectiveness to the new participative methodology for developing websites and partly reflect the differences between ISDM and website development methodologies.

6.4.1   Human Factors Methodology for Designing Websites

Vora (1998) describes a methodology which provides for the development of effective HCI for websites, with the main task being to have a clear understanding of user needs, with particular attention given to: the types of users and their characteristics; and their specific tasks and environments. Vora (1998) also focuses on other important issues in the framework: maintenance, evaluation (expert), and iterative testing (feedback).

This methodology focuses on the human interaction perspective in designing a website. It is basically a linear model where each stage must be completed before the next stage can be started. The stages are as follows:

- **Planning**: The designer needs to answer the following question "Why design a Website?" (Vora 1998, p. 155). The stage has several steps: defining the goals; identifying content owners and authors; understanding the users and environments; and finally, the most important aspect is to understand very precisely the users' needs.
- **Analysis**: During this stage, "decisions are made about both content and process" (Vora 1998, p. 156). "Content" refers to the material necessary to meet identified user tasks, addressing the information needs. The "process" refers to how the information should be maintained and how "interactive aspects of the websites are handled behind the scenes so that they are transparent to users" (Vora 1998, pp. 156–157).
- **Design and Development**: "information gathered in the earlier stages is translated into actual design" (Vora 1998, p. 160)
- **Usability Testing**: The key to a successful system or (Website) is iterative testing. This testing should occur not only in the final stage, but also in every stage to ensure that the system is on the correct track.
- **Implementation**: This stage is very practical and straightforward, as the designer will transfer the system (or website) to a specific location, to be used by the real user.

- **Maintenance**: This stage is very important. The designer and content providers need to provide up-to-date information on the site to make sure that the changes meet the user needs and to make the site more interesting and useful for the users.

However, this methodology does not specify user participation except in testing and planning. Users can also play a key role in defining content. According to Mayhew, these concepts are very important, especially from the users' perspective, as "One of its great weaknesses, is its lack of quality control for both the content and for presentation" (Mayhew 1998, p. 2). Furthermore, a procedure for addressing user disabilities was missing in Vora's methodology as "designers should keep in mind that the target population includes millions of potential users of web pages who have various handicapping sensory and physical conditions" (Laux 1998, p. 87). Table 6.6 shows that usability and iteration are the main aspects available in the Human Factor Methodology for Designing Websites. Usability is a very important aspect in this methodology with moderate to maximum rating to ensure that the website meets users' requirements in respect of performance and satisfaction. Iteration is available with minimum and moderate ratings in most stages, to ensure that the system is on the correct track. With respect to user participation, it is available only in the planning, testing, and evaluation stages with minimum rating, to identify user goals and understand their environments, and to test the product and make sure it meets users' desires. Finally, the real interaction is available only in the analysis and maintenance stages with moderate to maximum rating to ensure that the website has met users' requirements and—the most important aspect—to make it attractive and approachable to the users.

In the Human Factor Methodology for Designing Websites, there are five strong stages: planning, analysis, testing, evaluation, and maintenance. Planning and analysis are essential stages. The former will define the users' goals and examine the environment very carefully in order to meet the users' needs. The main areas of focus of the analysis stage are content (materials to suit user tasks) and process (how information should be maintained). In this methodology, the testing stage is iterative involving "expert evaluation," which means experts will evaluate the website and suggest solutions to problems. Finally, the maintenance stage is also important in this methodology. To make the website more interesting and to attract more users to visit it, designer and content providers need to provide up-to-date information in the site.

6.4.2   Relationship Management Methodology (RMM)

Isakowitz et al. (1995) describe a methodology, which provides for the development of effective websites for highly structured applications such as online conference proceedings, directories, academic journals, courseware, and electronic commerce.

In other words, this methodology is "most suited to applications that have a regular structure, especially where there is a frequent need to update the information to keep

Table 6.6 Human factor methodology for designing websites (HFMDW)—prepared by Tomayess Issa

Stages Principles	Planning	Analysis	Design	Testing	Implementation	Evaluation	Maintenance	Extra stages
User participation	1	0	0	1	0	1	0	Usability goals development
Usability	2	3	1	3	0	3	0	
Iteration	1	1	1	2	1	2	1	
Real interaction	0	2	0	0	0	0	3	
Strongest stage in HFMDW	☑	☑	–	☑	–	☑	☑	

the system current" (Isakowitz et al. 1995, p. 43). The main goal of this methodology is to reduce complexity and make the website easy to navigate and maintain, thereby saving time and money, and making it more attractive to the users. This methodology is divided into four sequential stages, where each stage must be completed before the next can be started. The feedback loops between the RMM design stages are shown by dashed lines, while the remaining feedback loops, "although present in RMM, are not shown" (Isakowitz et al. 1995, p. 39).

The stages of RMM may be described as follows:

- **Feasibility**: This stage provides the foundation for the RMM design methodology, as via this stage, the designer needs to define the objectives, user requirements, user analysis, and cost-benefit analysis.
- **Hardware Selection**: This stage involves definition of the hardware requirements for the website.
- **Information/Navigation Requirements Analysis**: During this stage, the designer identifies user tasks and develops an understanding of the information needs and likely use scenarios.
- **Design Methodology**: This stage provides the foundation for designing the relationship between the entities in the website. The stage has several steps, such as E-R design, entity design, navigation design, conversion protocol design, user interface screen design, and run-time behavior design.
- *E-R Design (S1)*: This step of the design process "represents a study of the relevant entities and relationships of the application domain" (Isakowitz et al. 1995, p. 39). These entities and relationship of data are considered the basis for the hypermedia applications.
- *Entity Design (S2)*: This step is unique to the hypermedia application, as, through it, the designer will determine "how the information in the chosen entities will be presented to users and how they may access it" (Isakowitz et al. 1995, p. 40).
- *Navigation Design (S3)*: This step defines how the navigation will be established between the entities, which are based on "associative relationships" (Isakowitz et al. 1995, p. 41).
- *Conversion Protocol Design (S4)*: This step sets the conversion rules to "transform each element of the RMDM diagram into an object in the target platform" (Isakowitz et al. 1995, p. 43).
- *User Interface Design (S5)*: This step involves the design of screen layouts for each object appearing in the RMDM diagram obtained in Step 3. Via this step, the designer will design the "button layouts, appearance of nodes and indices and location of navigational aids" (Isakowitz et al. 1995, p. 43).
- *Run-Time Behavior Design (S6)*: This step considers the "volatility and the size of the domain to decide whether node contents and link endpoints are to be built during application development or dynamically computed on demand at runtime" (Isakowitz et al. 1995, p. 43).
- *Construction and Testing (Evaluation) (S7)*: This stage is similar to the one in the traditional software development process. Special care must be taken in this stage

to test the website to determine if it is running according to the user requirements, especially the navigational paths.

This methodology is best suited to large websites focusing on product catalogs and hypermedia front ends of databases. The main flaw of this methodology is that it is missing the maintenance stage. This concept is very valuable, particularly from the users' perception to attract new users to visit the website and to encourage the current users to visit and work with it. Finally, this methodology does not distinguish "between how information is abstracted and how it is presented. Relationships are just translated to navigational paths and no other communication among the entities is allowed" (Isakowitz et al. 2000). Iteration is available in the design stage with a moderate rating but in the rest of the stages with a minimum rating. The purpose of the iteration stage is to ensure that the website is running according to the user requirements, especially the navigational paths. To prevent any confusion in this methodology, the feedback loops in the design stage were shown as dashed lines, while the remaining feedback present in this methodology is not shown as in the diagram.

There are zero availability ratings for user participation, usability, and real interaction in this methodology. This means that these aspects are not well considered in this methodology.

The strongest stages in the RMM are the planning and design. The planning stage defines the objectives, user requirements and analysis, and cost-benefit analysis, while the design stage is the dominant stage in this methodology as the designer will classify: (1) the relationship between the entities in the website; (2) the navigational path between the entities; and (3) the design of screen and button layouts (see Table 6.7).

6.4.3 The W3DT Design Methodology

Bichler et al. 1996) describe the World Wide Web Design Technique (W3DT), a methodology especially for designing a large-scale web-based hypermedia application. This methodology focuses on two main parts: modeling techniques and computer-based design. The former gives the designer the possibility to "generate a running prototype of the system, including HTML-pages and CGI-scripts," while the latter allows the designer to define and draw a "graphical representation of a website's structure" (Bichler et al. 1996, p. 328). The major requirement for dealing with W3DT is to keep the models "clear and intuitively comprehensible" (Bichler et al. 1996, p. 328).

The essential design primitives and their interaction are best described by the W3DT meta model, which shows "the class hierarchy of the different elements" (Bichler et al. 1996, p. 330). The first essential design primitive is site. One or more diagrams can be found under the site, and each diagram serves two purposes:

Table 6.7 Relationship management methodology (RMM)—prepared by Tomayess Issa

Stages / Principles	Planning	Analysis	Design	Testing	Implementation	Evaluation	Maintenance	Extra stages
User participation	0	0	0	0	0	0	0	Hardware selection; navigation design; and user interface
Usability	0	0	0	0	0	0	0	
Iteration	1	1	2	1	1	1	1	
Real interaction	0	0	0	0	0	0	0	
Strongest stage in RMM	☑	–	☑	–	–	–	–	

to indicate a hierarchical refinement of a model and to include sub-models into a unified view (Bichler et al. 1996, p. 330).

Usually, a diagram consists of one page with the option to have "layout" and "link" on the same page. The main purpose of layout is to hold information about website headers, footers, and background images. On the other hand, the link can be more than just a "hypertext reference to another document" (Bichler et al. 1996, p. 330). Furthermore, page, form, index, and menu are the basic elements for building a "hypermedia application information domain" (Bichler et al. 1996, p. 330). There is no major difference between an index and a menu in the W3DT meta model, as the former is used to list a complete set of links, while the latter is a "navigational aid with the main purpose to provide access structures" (Bichler et al. 1996, p. 330). It was noted that this methodology has been widely used by several groups of students at universities, colleges, and website developers in organizations "showing very promising results" (Bichler et al. 1996, p. 333). However, this methodology is missing seven essential concepts: planning, analysis, implementation, testing, iteration, evaluation, and maintenance. These stages are very important in the development process as, via them, the designer will test and evaluate the system (or the website) to check whether users' requirements were met.

Table 6.8 indicates zero ratings for the four key principles in the W3DT Design Methodology. This means that none of the above four key principles were incorporated in this methodology to any significant degree. The strongest stage in the W3DT Design Methodology is the design stage. This stage gives the designer the chance: (1) to generate a first trial product of the system with a hypermedia application and (2) to draw a graphical representation of the website construction.

6.4.4 Information Development Methodology for the Web

John December (1996) describes a methodology which provides for the development of effective websites for technical communicators, writers, designers, and software developers. The main task of this methodology is to decrease difficulty and make the website easy to navigate and maintain, and more attractive to the users. This methodology is very usable for dynamic and competitive web design. December argued that this "methodology was based on the characteristics and qualities of the web on the experiences of web users" (December 1996, p. 372). This methodology is divided into six sequential stages (or elements, according to John December), where each must be completed before the next stage can be started. The stages are as follows:

- **Planning for the Audience and Purpose**: This stage defines several items, which are very useful to build a website, such as the purpose of the website and audience information. The audience information can include concerns, background, and characteristics. December stated that this planning and analysis require asking

Table 6.8 W3DT Design Methodology—prepared by Tomayess Issa

Stages Principles	Planning	Analysis	Design	Testing	Implementation	Evaluation	Maintenance	Extra stages
User participation	o	o	o	o	o	o	o	Navigation design and building a hypermedia application
Usability	o	o	o	o	o	o	o	
Iteration	o	o	o	o	o	o	o	
Real interaction	o	o	o	o	o	o	o	
Strongest stage in W3DT	–	–	☑	–	–	–	–	

and answering questions such as "Who will use this web? And what will they gain from it?" (December 2003)

- **Setting Objectives and Gathering Domain Information**: After considering the purpose and audience, the designers and analysts need to concentrate on the objectives and goals that the website needs to accomplish.
- **Designing a Web**: To make the web flexible, efficient, and easy to use a relationship should be established between the pages of the web. Therefore, to design a website, the designer should have a thorough grounding in "hypertext, multimedia, Java and other programming possibilities as well as knowledge about how particular web structures affect an audience" (December 2003).
- **Implementing a Web**: The purpose behind this stage is to create files of HTML and other software. The initial implementation might be a "prototype which is not released publicly, but available for analysis [and use] by a set of representative users" (December 2003).
- **Analyzing a Web**: This stage involves the designer examining the web structure and contents to determine if it meets the objectives, goals, and the purpose of the web.
- **The Web's release and promotion and ongoing innovation** involves the web being "publicly released for general web audiences, potential users and current users" (December 1996, p. 372). Furthermore, it involves ongoing support and work to improve the web in order to meet the user requirements.

This methodology is limited to websites for information, art, general services, and entertainment. The methodology is missing two essential aspects: iteration and evaluation stages. These concepts are very important, especially from the users' perspective. Table 6.9 indicated that the four key principles have zero ratings in the Information Development Methodology for the Web except for user participation and real interaction, which have a minimal rating in the implementation stage because of the role of representative users in reviewing the prototype.

The real interaction is available in the maintenance stage to improve the web in order to meet the user needs. The strongest stage in Information Development Methodology for the web is implementation. This stage releases the first sketch of the website and is checked by representative users in order to make sure it complies with the user requirements.

6.4.5 The Web Site Design Method (WSDM)

Olga De Troyer (1998) describes a methodology for website design. The main goal for this new methodology is to develop a site which provides information "in such a way that both the provider and the inquirer benefit from it" (De Troyer and Leune 1998, p. 88). The main mission statement for this methodology is [to describe] the subject purpose and the target audience for this website. Without giving good consideration

Table 6.9 Information development methodology for the web—prepared by Tomayess Issa

Stages Principles	Planning	Analysis	Design	Testing	Implementation	Evaluation	Maintenance	Extra stages
User participation	0	0	0	0	1	0	0	Promotion and prototyping (is available under the implementation phase)
Usability	0	0	0	0	0	0	0	
Iteration	0	0	0	0	0	0	0	
Real interaction	0	0	0	0	0	0	1	
Strongest stage in IDMW	–	–	–	–	☑	–	–	

to the mission statement, there "is no proper basis for decision making or for the evaluation of the effectiveness of the website" (De Troyer 1998, p. 53).

This methodology has adopted the "user-centered" approach in order to create effective communication and to define the different types of users and characteristics and their information requirements. This will lead to definition of the "perspectives." A perspective "is a kind of user subclass," which means, "all users in a user class with the same characteristics and usability requirements" (De Troyer 1998, pp. 54–55). This methodology consists of the following stages: user modeling, conceptual design, implementation design, and the actual implementation.

- **User Modeling**: This stage is divided into two steps: user classification and user class description. The purpose behind this stage is to concentrate "on the potential users of the website" (De Troyer et al. 1998, p. 88).

  - *User Classification*: This step will help the designers to identify the future users or visitors of the website and classify them into user classes. Therefore, the purpose of this step is to identify the target audience by "looking at the organization or the business process which the website should support" (De Troyer 1998, p. 53).
  - *User Class Description*: This step will help the designer to analyze in more detail the user types in order to identify not only their "information requirements but also their usability requirements and characteristics" (De Troyer 1998, p. 54). Examples of information requirements are "levels of experience with websites in general, language issues, education/intellectual abilities, age." Some of this information can be "translated into usability requirements" (De Troyer 1998, p. 54).

- **User Conceptual Design:** This stage is divided into two steps: user modeling and the navigational design. This stage utilizes different "user classes and their perspectives" which will allow the users to efficiently "navigate through the website" as each user class has its own "navigation track" (De Troyer et al. 1998, p. 90).

  - Object Modeling: This step will help the designers to identify information requirements of different user classes and their perspective.
  - Navigational Design: This defines the specific navigation path through the website for each user class.

- **The Implementation Design:** This stage will help the designer to design the "look and feel" of the website, to "create a consistent, pleasing and efficient look and feel for the conceptual design made in the previous phase" (De Troyer 1998, p. 55).
- **The implementation** is the "actual realization of the website using the chosen implementation environment, e.g., HTML" (De Troyer 1998, p. 55).

The WSDM methodology is "user centered" rather than "data driven," which means the starting, point for this methodology "is the set of potential visitors of the Website" (De Troyer et al. 1998, p. 85). The user participation is not strong in this methodology; however, the WSDM methodology seeks to learn more information about the users in respect of their knowledge in dealing with the website, language, education, and age. This information will help the designer to translate these user characteristics into usability needs and requirements of the website. However, the WSDM methodology is missing a few stages in the development process, namely: testing, iteration, evaluation, and maintenance. These stages are important, as, through them, the designer will learn if the website meets users' requirements.

Table 6.10 indicates that user participation is covered in the planning, analysis, and design stages with minimal rating, as the designer is seeking to gain more general information about the users such as language, age, and education, as some of this information will be translated into usability requirements. Usability aspects are available in planning, analysis, design, and implementation with a moderate rating, while the real interaction has a similar rating but in analysis and design. For iteration, the rating is zero, which means it is not considered in this methodology. The strongest stages in the WSDM are the planning, analysis, and design. The planning stage will help the designer to identify the target audience to the website and to classify them into user classes, while the analysis stage will help the designer to analyze in more detail the user types in order to identify information and usability requirements and characteristics. Finally, the design stage will help the designers to identify the information required, how it will be presented, and the navigation paths for user types.

6.4.6 Summary of Methodologies for Developing Websites

This section will provide a summary behind the methodologies for developing Websites:

For example, in the **Human Factor Methodology for Designing Websites,** the four key principles are available but in varying degrees in different stages. Usability is very dominant in analysis, testing, and evaluation stages with maximum rating, while in the planning and design stages it has a moderate rating. This means that usability is a very significant aspect in this methodology to ensure that the website is running without any errors and enhancing job satisfaction. Iteration is available in some stages with minimum rating that is in planning, analysis, design, implementation, and maintenance, with a moderate rating in testing. User participation is available only in the planning, testing, and evaluation stages with a minimum rating, while the real interaction has a moderate rating in analysis and maximum rating in the maintenance stage. In the Human Factor Methodology for Designing Websites, there are five strongest stages: planning, analysis, testing, evaluation, and maintenance. Planning and analysis are essential stages for defining the users' goals, understanding the environment, and the way that information should be maintained. The testing and evaluation stages are also very important. Finally, the maintenance stage incorporates

Table 6.10 Web site design method (WSDM)—prepared by Tomayess Issa

Stages / Principles	Planning	Analysis	Design	Testing	Implementation	Evaluation	Maintenance	Extra stages
User participation	1	1	1	0	0	0	0	User modeling and conceptual design
Usability	2	2	2	0	1	0	0	
Iteration	0	0	0	0	0	0	0	
Real interaction	0	2	2	0	0	0	0	
Strongest stage in WSDM	☑	☑	☑	–	–	–	–	

the provision of up-to-date information, in order to make the website more attractive and interesting.

In the **Relationship Management Methodology (RMM)**, only iteration is available with minimum or moderate ratings in all the stages. Zero rating for user participation usability and real interaction in this methodology means that usability, user participation, and real interaction are largely ignored. The strongest stages in the RMM are design and planning. Design and planning are essential, as the former will help the designer to define the relationship and navigational path between the entities and to design the screen and button layouts;, while the latter will define users' goals and an understanding of the cost-benefit analysis.

The four key principles have zero ratings in **the W3DT Design Methodology** and the **Information Development Methodology for the Web** except for a minimum rating for user participation in the implementation stage and with minimum rating for real interaction in the maintenance of the latter methodology. This means that the four key principles are largely ignored in these methodologies. The strongest stage in the **W3DT Design Methodology** is the design stage. The strongest stage in the **Information Development Methodology for the Web** is implementation. This stage permits the users to check the first draft of the website to ensure it meets the users' requirements and needs.

Finally, the four key principles are addressed in the **Web Site Design Method (WSDM),** except for iteration. User participation is incorporated into various stages, such as in planning, analysis, and design with minimum rating, while usability is available with minimum and moderate rating in planning, analysis, implementation, and design, respectively, and real interaction is available with moderate ratings in the analysis and design. The strongest stage in WSDM is the design stage. This stage will help the designers to distinguish the future users or visitors of the website and gain more information about their characteristics.

After reviewing the methodologies for developing websites, extra stages are collected from these methodologies (see Table 6.11). The main focuses of these extra stages are usability, navigation, promotion, prototyping, and identifying user types. These stages are very significant for developing websites. Therefore, most of

Table 6.11 Extra stages from methodologies for developing Websites—prepared by Tomayess Issa

Methodology (developing Websites)	Extra stages
Human Factor Methodology for Designing Websites	Usability goals development
Relationship Management Methodology (RMM)	Hardware selection; navigation design; and user interface
The W3DT Design Methodology	Navigation design and building a hypermedia application
Information Development Methodology for the Web	Promotion and prototyping "is available under the implementation phase"
The Web Site Design Method (WSDM)	User modeling and conceptual design

these stages will be taken into consideration by the researcher to be added to the new participative framework for developing websites.

Table 6.12 demonstrates the strongest stages from methodologies for developing websites and presents the rating availability for the four key principles in each stage. It was noticed that all the stages were covered in the methodologies for development of websites as the main focus for these methodologies is:

- Defining the users' goals and understanding the environment very precisely in order to meet the users' needs and analyze the cost benefits.
- Defining the materials to identify user tasks and how information should be maintained.
- Defining the navigational path between the entities on the website, designing of screen and button layouts, generating a first trial product of the system, and defining user usability requirements and their characteristics.
- Releasing the first sketch of the website that will be checked by representative users in order to ensure that it complies with the user requirements.
- Making the website more interesting and attractive so that more users visit it, via content providers contributing up-to-date information to the site.

6.5 Marketing Methodologies

This section will examine the actual values added by marketing methodologies and the benefits they will bring to the e-commerce framework, especially in developing websites. In this section, the researcher will examine several methodologies from the marketing perspective such as E-Marketing Plan, and will review methodologies which were created by companies, which are developing websites for marketing. At the end of each methodology section, the researcher will present a table showing: (1) how the four key principles are addressed in each stage within the methodology; (2) the strongest stage for each methodology for developing websites; and (3) the extra stages of each methodology. These extra stages will help the researcher to develop a more comprehensive structure for the new participative methodology for developing marketing websites.

6.5.1 E-Marketing Plan

The E-Marketing Plan is a "guiding, dynamic document that links the firm's e-business strategy with technology-driven marketing strategies and lays out details for plan implementation through marketing management" (Strauss et al. 2006, p. 46). The main ideas behind an E-Marketing Plan are: (1) to achieve an effective and efficient e-business objective; (2) to increase revenues and reduce costs; and (3) to serve "as a roadmap to guide the direction of the firm, allocate resources, and make tough decisions at critical junctures" (Strauss et al. 2003).

Table 6.12 Summary of strongest stages from methodologies for developing Websites—prepared by Tomayess Issa

Stage	Methodologies for developing Websites	Principles			
		User participation	Usability	Iteration	Real interaction
Planning	Human Factor Methodology for Designing Websites (HFMDW)	1	2	1	0
	Relationship Management Methodology (RMM)	0	0	1	0
	The Web Site Design Method (WSDM)	1	2	0	0
Analysis	Human Factor Methodology for Designing Websites (HFMDW)	0	3	1	2
	The Web Site Design Method (WSDM)	1	2	0	2
Design	Relationship Management Methodology (RMM)	0	0	2	0
	The W3DT Design Methodology	0	0	0	0
	The Web Site Design Method (WSDM)	1	2	0	2
Testing	Human Factor Methodology for Designing Websites (HFMDW)	1	3	2	0
Implementation	Information Development Methodology for the Web	1	0	0	0

(continued)

Table 6.12 (continued)

Stage	Methodologies for developing Websites	Principles			
		User participation	Usability	Iteration	Real interaction
Evaluation	Human Factor Methodology for Designing Websites (HFMDW)	0	3	2	0
Maintenance	Human Factor Methodology for Designing Websites (HFMDW)	0	0	1	3
	Information Development Methodology for the Web	0	0	0	1

Strauss et al. (2003) suggest that there are two common types of E-Marketing Plans: the "napkin plan" and the "venture capital plan." The former approach is to just "jot ideas on a napkin over lunch or cocktails and then run off to find financing" (Strauss et al. 2006, p. 47), and however, these plans work only sometimes, while the latter plan basically focuses on building a suitable business plan to increase the profit and reduce the cost. Therefore, the traditional marketing plan needs to be introduced to define and clarify key questions about topics such as capital, new customers, product and service, pricing and customer support required to retain the customers. Sound planning and "thoughtful implementation is needed for long-term success in business" (Strauss et al. 2003).

The E-Marketing Plan is divided into seven steps:

- **Situation Analysis:** This step will help the marketers to define and review the firm's environment and involves strengths, weakness, opportunities, and threats (SWOT) analyses. Strengths and weakness of the company's internal situation need to be identified, new opportunities need to be defined to improve the current situation of the company, while the threats "are areas of exposure" (Strauss et al. 2006, p. 50). Also under this step, a review and analysis of the existing marketing plan needs to be carried out to identify appropriate strategies, objectives, and performance metrics for e-business.
- **E-Marketing Strategic Planning:** This step involves "determining the fit between the organization's objectives, skills and resources and its changing market opportunities" (Strauss et al. 2006, p. 51). Additionally, the marketers will create a sustainable e-marketing strategy for the e-business goals from "marketers design segmentation, targeting, differentiation, and positioning strategies" (Strauss et al. 2003). This includes demographics, geographic location, psychographics, and

behavior of potential customers. This information will help the marketers to formulate the e-marketing objectives.

- **Objectives**: Three main issues need to be defined in an E-Marketing Plan: task (what one is planning to achieve by building this e-business); measurable quantity (how much); and time frame (setting a time to accomplish the e-business job).
- **E-Marketing Strategies**: In this step, the marketers need to identify the 4Ps (product, pricing place, and promotion) and the relationship management requirements to "achieve plan objectives regarding the offer" (Strauss et al. 2006, p. 53)— product: what is planned to be produced at the end (by building the e-business) in terms of service, information, selling products, or advertising; pricing: what it will cost for the e-business to be implemented; place: the location of the e-business work; and promotion: the techniques that will need to be adopted in order to promote the e-business work. The relationship management strategies need to identify how to "build relationships with a firm's partners, supply chain members, or customers" (Strauss et al. 2006, p. 57). Some companies use customer relationship management (CRM) or partner relationship management (PRM) approaches. PRM software is used to build and develop a complete database, which retains information about business partner capabilities and communication, while the purpose of the CRM software is "to retain customers and increase average order values and lifetime value" (Strauss et al. 2006, p. 57).
- **Implementation Plan**: The marketers select the 4Ps, relationship management strategies, and other tactics to achieve the e-marketing objectives and to develop the implementation plan. To achieve the implementation plan, the firm needs to check if the following aspects are available to accomplish the firm's objectives "staff, department structure, application service providers, and other outside firms" (Strauss et al. 2006, p. 57). Furthermore, special tactics will be used on the website to collect information about users who are dealing with it, such as forms, feedback e-mail, and online surveys. According to Strauss et al. (2003), additional tactics, which can be used to collect information, include: "(1) Website log analysis software helps firms review user behavior at the site and make changes to better meet the needs of users, (2) Business intelligence uses the Internet for secondary research, assisting firms in understanding competitors and other market forces."
- **Budget:** The key aspect of this stage is to identify the expected costs and returns from the investment. Returns are matched "against costs to develop a cost/benefit analysis, ROI calculation, or internal rate of return (IRR)" (Strauss et al. 2003) to determine if it is worthwhile to continue with the project. Furthermore, during the implementation stage, the marketers observe whether the results (cost and revenue) are on the correct track for achieving the predicted cost/benefit ratio.
- **Evaluation plan** is used to evaluate the success of the website. The tracking system should be available before activating the website. "E-marketers use tracking systems to measure results and evaluate the plan's success on a continuous basis" (Strauss et al. 2006, p. 60).

This E-Marketing Plan is a very important tactic for the marketers to gain more information about the current situation of the business before releasing the new version of e-business. However, this plan lacks a few stages which need to be available in order to achieve user exceptions and requirements, such as design, testing, iteration, and maintenance.

The strongest stages in the E-Marketing Plan are e-marketing strategies (under the planning stage), the implementation stage, and the evaluation stage. E-marketing strategies will allow the designer to identify the 4Ps: product, pricing, place, and promotion, and the relationship management requirements to achieve plan objectives for the website. In the implementation stage, the marketers will utilize the 4Ps, the relationship management strategies, and other tactics to achieve the e-marketing objectives. The evaluation stage involves tracking systems to measure results and evaluate the plans for the website.

Table 6.13 indicates that usability and iteration have zero ratings for this methodology. User participation is available in the planning and implementation stages with minimal rating, and real interaction is available in the evaluation stage with maximum rating. To formulate the e-marking objectives, the marketers will collect general information about the users such as demographics, geographic location, psychographics, and behavior of potential users in the planning stage, while in the implementation stage, special tactics will be used to collect information about the users such as forms, feedback e-mail, and online surveys.

6.5.2 The Advertures Company Methodology

The Advertures Company released a process methodology to enhance the development of websites from a marketing perspective in 2004. This methodology has five stages, each of which should be completed before moving to the next stage.

- **Orientate**: This stage will help the designers to know why they are developing this website. In this stage, the designer will define the following concepts: the goals, product details, and competition. These concepts will also help to determine the cost and time for establishing this website.
- **Blueprint:** This stage will produce the first sketch for the website, where the "marketing, technology and creativity collide; banging heads and eventually coming upon the best way to mix all three aspects and create the optimum product" (Advertures 2004).
- **Model:** This stage will combine the technology possibilities and the creativity from the sketch to produce the working model.
- **Build:** During this stage, the designers will build up the new system and make sure that the proposed website is tested repeatedly until it meets users' requirements.
- **Maintain:** Through this stage, the website will be maintained in order to "continue functioning at optimum levels" (Advertures 2004).

Table 6.13 E-Marketing Plan—prepared by Tomayess Issa

Stages Principles	Planning	Analysis	Design	Testing	Implementation	Evaluation	Maintenance	Extra stages
User participation	1	0	0	0	1	0	0	E-marketing strategies
Usability	0	0	0	0	0	0	0	Objectives and budget
Iteration	0	0	0	0	0	0	0	
Real interaction	0	0	0	0	0	3	0	
Strongest stage in E-Marketing Plan	☑	–	–	–	☑	☑	–	

From the Advertures Company point of view, this methodology will meet the users' requirements when building a website from the marketing perspective; however, not all the possible stages are available in this methodology. When compared with other system development processes, it lacks detailed design.

Table 6.14 shows that user participation; usability and real interaction have zero rating, while iteration is available in the testing stage with moderate rating to ensure that the website is tested repeatedly until it meets users' requirements. The strongest stage in this methodology is testing, which allows the designer to test the project repeatedly until it meets users' requests and desires.

6.5.3 The Market-Vantage (Internet Performance Marketing) Methodology

The Market-Vantage Company introduced a new methodology process for developing websites to enhance the strategy of the websites in order to "reduce cost, increase customer loyalty and market analysis" (Market-Vantage 2003). This methodology has four stages, each of which should be completed before moving to the next stage.

- **Internet Marketing Goals, Target Markets, and Strategy**: This stage helps the designers to ask the following questions in the planning process: What are you selling? Who are the buyers? Who are your competitors? In addition, how can potential customers find the product? (Market-Vantage 2003). Answers to these questions will give the designer a full picture of the purpose behind building this website.
- **Define/Refine Internet Marketing Strategy**: This stage helps the designers in two aspects: learning about users [the purpose behind the visit and tracking their visit] and how the business will be enhanced by using the Internet for introducing the new products.
- **Implementation**: This stage establishes the website so that the users can start using the new product and check if it meets their requirements.
- **Measurement** is part of ongoing maintenance of the website and checking if the results of using the website are meeting its goals, using software to track current and new users. Continuing support and recommendations are available from the designer to the website manager.

We notice that this methodology includes iteration, so as to ensure that the website is meeting the user requirements and providing appropriate company outcomes. However, this methodology is missing a few stages such as detailed analysis and design. These stages are imperative in developing a website so that the website achieves the goals of e-business as well as meeting users' requirements.

Table 6.15 identifies that user participation and usability have zero rating (except in the maintenance stage), while iteration is available in the implementation stage

Table 6.14 Advertures Company Methodology—prepared by Tomayess Issa

Stages Principles	Planning	Analysis	Design	Testing	Implementation	Evaluation	Maintenance	Extra stages
User participation	0	0	0	0	0	0	0	Blueprint and model
Usability	0	0	0	0	0	0	0	
Iteration	0	0	0	2	0	0	0	
Real interaction	0	0	0	0	0	0	0	
Strongest stage in the Advertures Company Methodology	–	–	–	☑	–	–	–	

Table 6.15 Market-Vantage (Internet Performance Marketing) Methodology—prepared by Tomayess Issa

Stages Principles	Planning	Analysis	Design	Testing	Implementation	Evaluation	Maintenance	Extra stages
User participation	1	0	0	0	0	0	2	Define/refine Internet marketing strategy and measurement
Usability	0	0	0	0	0	0	0	
Iteration	0	0	0	0	2	0	0	
Real interaction	0	0	0	0	0	0	2	
Strongest stage in the Market-Vantage methodology	☑	–	–	–	☑	–	☑	

with moderate rating to ensure that the website meets users' requirements. Real interaction is available in the maintenance stage with moderate rating to check if the website meets users' requirements and needs after the changes have been made.

The strongest stages in the Market-Vantage (Internet Performance Marketing) Methodology are the planning, implementation, and the maintenance stage, which is under the measurement stage. The planning stage will help the designers to identify the purpose behind building the website, namely the products/service being sold, the firms' competitors and buyers, and how to find the product via the web. The implementation stage is important in Market-Vantage to allow users to use the new product and to check if it meets their requirements. User information is used in the maintenance stage to review ongoing performance of the website.

6.5.4 EnSky's Unique Methodology

EnSky Company initiated a new methodology for developing websites from the marketing perspective. This methodology has nine stages, each of which should be completed before moving to the next stage.

- **Evaluation Overview:** This methodology divides the evaluation aspect into two types: pre- and post-evaluation. The former is a phase to define the user needs and requirements for success and to determine the approach to be used in the latter stages, namely, to define "the methods to track the results in post-evaluation" (EnSky 1997). The initial pre-evaluation stage establishes the goals of the project and identifies the existing branding, "marketing strategies, middle market demographics, competitors, and developing an understanding of the business and sales models" (EnSky 1997). According to EnSky's methodology, the post-evaluation process is very useful to measure the effectiveness of the site against the goals, which were set in the pre-evaluation.
- **Design:** During this stage, the designer will define the specifications and requirements and document the design of the look of the "end product that extends from the branding and marketing strategies already employed" (EnSky 1997).
- **Develop:** This stage will carry out the outcomes from the design phase to build the website by using various tools such as templates and graphical files, which were created in the design stage.
- **Testing:** During this stage, the prototype website will be tested to determine if it meets the requirements of the users. According to the EnSky methodology, once the "testing requirements have been met and approved by the client the project is ready for deployment" (EnSky 1997).
- **Deployment:** During this stage, the designer will transfer all the files of the website to the in-house web server. After this stage, the designer will follow the methodology by using the promotion and maintenance stages so as to begin "the process of both updating the content on the site to keep it relevant and marketing

the site to create awareness and drive traffic to it ensuring ultimate ROI" (EnSky 1997).

- **Promote:** This stage will help to promote the website to the public, by using various tools such as press releases, link building, banner ad campaigns, and paid search engine or directory listing campaigns. These processes will be repeated from time to time in order to make sure that the promoting phase is effective.
- **Maintain:** Via this stage, the designer will make sure that the website is updated and maintained regularly and facilitates "the adoption of global technological advances" (EnSky 1997).
- **ROI:** This stage reviews the cost and investment of developing the website and compares it with likely returns.
- **Measurement** is part of the ongoing maintenance of the website and is integral in determining the ROI. According to EnSky, various types of tools are used for these measurements such as "search engine ranking and website visitor statistics, tracking sales, new customers etc." (EnSky 1997).

This methodology contains most of the stages, which are needed for the designer to develop a website which meets the e-business objectives and to evaluate the returns against the costs. However, two stages are missing—detailed analysis and iteration.

Table 6.16 indicates zero rating for the four key principles except for minimal user participation in the testing and maintenance stages and a minimal rating for real interaction in the maintenance stage. This means that the four key principles are mainly ignored in this methodology. The strongest stage is maintenance (under the measurement stage). This stage is important to the designer and users simultaneously, as it will attract more users to visit the site. In addition, this stage includes changes and correction of errors in hardware and software to meet user requirements.

6.5.5 Review of Marketing Methodologies

The analysis above indicates that most stages in the marketing methodologies are similar to those in lifecycles, methodologies, and models, with extra stages focusing on the marketing perspective, such as measurement, promotion, and cost/benefit analysis. These extra stages will help the firm to achieve "its desired results as measured by performance metrics according to the specifications of the e-business model and e-business strategy" (Strauss et al. 2006, p. 60).

6.5.6 Summary of Marketing Methodologies

This section will provide a summary behind the marketing methodologies.

E-Marketing Plan usability and iteration have zero rating, while user participation is available in planning and implementation with minimal rating to collect

Table 6.16 EnSky's unique methodology—prepared by Tomayess Issa

Stages / Principles	Planning	Analysis	Design	Testing	Implementation	Evaluation	Maintenance	Extra stages
User participation	0	0	0	1	0	0	1	Develop; ROI; measurement; and promotion
Usability	0	0	0	0	0	0	0	
Iteration	0	0	0	0	0	0	0	
Real interaction	0	0	0	0	0	0	1	
Strongest stage in EnSky's unique methodology	–	–	–	–	–	–	☑	

general information about the users. Real interaction is available in the evaluation stage with maximum rating as the e-marketers use tracking systems to measure the results and ensure that the website meets users' requirements. The strongest stages in E-Marketing Plan are e-marketing strategies, implementation, and evaluation. E-marketing strategies will allow the designer to identify the 4Ps: product, pricing, place, and promotion, and the relationship management requirements to achieve plan objectives for the website. To achieve the implementation stage, the firm needs to check if all the objectives are available to accomplish the firm's needs. The evaluation stage is for tracking the users' behaviors to establish whether the website meets their requirements.

In **the Adventures Company Methodology,** user participation, usability, and real interaction have zero rating, while iteration is available in the testing stage with moderate rating. Testing is the strongest stage in this methodology as this allows the designer to test the project frequently until it meets users' requests and desires.

The **Market-Vantage (Internet Performance Marketing) Methodology** is similar to the Adventures Company Methodology, as user participation and usability have zero rating (except for a moderate rating for participation in the maintenance stage). Iteration can be found in the implementation stage to ensure that the website meets users' requirements. Real interaction is available in the maintenance stage. The strongest stages are planning, implementation, and maintenance. The planning stage will allow the designers to gain more information about the rationale behind building the website, i.e., what is being sold; the firm's competitors and buyers; and how to find the product via the web. The implementation stage will allow the users to use the new product and check if it meets their needs. User satisfaction is tested during the maintenance stage.

The **EnSky's Unique Methodology** has zero ratings for the four key principles, except for a minimal rating for participation in the testing stage and real interaction in the maintenance stage. The strongest stage in EnSky's Unique Methodology is maintenance. This stage involves ongoing changes and correction of errors in hardware and software, in order to continue to meet user requirements.

After reviewing the marketing methodologies, extra stages were identified (see Table 6.17), focusing mainly on: promotion, prototyping, budget, return on investment (ROI), and measurement. These stages are important for developing websites

Table 6.17 Extra stages of marketing methodologies—prepared by Tomayess Issa

Methodology (marketing)	Extra stages
E-Marketing Plan	E-marketing strategies, objectives, and budget
The Adventures Company Methodology	Blueprint and model
The Market-Vantage (Internet Performance Marketing) Methodology	Define/refine internet marketing strategy and measurement
EnSky's Unique Methodology	Develop, ROI, measurement, and promotion

from the marketing perspective. Therefore, the researcher will take into consideration these stages for the new participative framework for developing websites. The key techniques involved are:

- Identify the 4Ps for the E-Marketing Plan: product, pricing, place, and promotion.
- Identify the time frame to accomplish the job.
- Identify the expected returns from investment.
- Produce the first sketch for the website, evaluate it, and then move on to produce the working model.
- Learn about the users by tracking their visit and the purpose behind the visit.

Table 6.18 demonstrates the strongest stages for marketing methodologies and presents the rating for the four key principles in each stage. The main focuses of these methodologies are:

- Identify the product, pricing, place, promotion, and the relationship management requirements to achieve plan objectives for the website.
- Planning the purpose behind building the website, i.e., what are you selling; your competitors and buyers; and how to find the product via the web.
- Testing the website repeatedly until it meets users' requests and desires.
- Maintaining the website to attract more users (new as well as old) to visit it.

6.6   Detailed Website Design and Implementation

The previous sections highlighted the need for a detailed approach to website design. This can lead to an effective website implementation, including organizational aspects. Two types of approaches will be discussed from the web-based hypermedia application perspectives in this section: the object-oriented hypermedia design model and the implementation model.

6.6.1   The Object-Oriented Hypermedia Design Model (OOHDM)

Schwabe and Rossi (1995) describe an object-oriented hypermedia design model (OOHDM), a new model especially for designing a complex web-based hypermedia application. The main aims of this approach are to: reduce complexity, make the website easy to navigate and maintain, thereby saving time and money, and make it more attractive to the users. This approach clearly separates the "navigational from conceptual design by defining different modeling primitives in each step" (Schwabe and Rossi 1995, p. 46). This approach is divided into four sequential stages, where each must be completed before the next stage can be started, although iteration can

Table 6.18 Summary of marketing methodologies

Stage	Marketing methodologies	Principles			
		User participation	Usability	Iteration	Real interaction
Planning	E-Marketing Plan	1	0	0	0
	The Market-Vantage (Internet Performance Marketing) Methodology	1	0	0	0
Analysis	–	–	–	–	–
Design	–	–	–	–	–
Testing	The Adventures Company Methodology	0	0	2	0
Implementation	E-Marketing Plan	1	0	0	0
	The Market-Vantage (Internet Performance Marketing) Methodology	0	0	2	0
Evaluation	E-Marketing Plan	–	–	–	3
Maintenance	The Market-Vantage Methodology	2	0	0	2
	EnSky's Unique Methodology	1	0	0	1

be used. Each stage "focuses on a particular design concern, and an object-oriented model is built" (Schwabe and Rossi 1995, p. 45). The stages are as follows:

- **Domain Analysis**: In this, stage the "conceptual model of the application domain is built using well-known object-oriented modeling principles" (Schwabe and Rossi 1995, p. 45).
- **Navigational Design**: In this stage, the navigational structure for the hypermedia application will be defined in "terms of navigational contexts (focusing on the users and their tasks), which are induced from navigation classes such as nodes, links, indices, and guided tours" (Schwabe and Rossi 1995, p. 46).
- **Abstract Interface Design**: This stage provides the "perceptible objects" (i.e., picture, a city map. etc.) in "terms of interface classes" (i.e., text fields and

Table 6.19   OOHDM methodology—extra stage—prepared by Tomayess Issa

The OOHDM methodology—extra stage
Design: Two aspects will be defined in this stage: (1) navigational design and (2) abstract interface design. The latter will define the navigational structure for the hypermedia application, while the former will establish the communication between the interface and navigation in the hypermedia application **Construction (implementation)**: involves the technical implementation of the design

buttons) (Schwabe and Rossi 1995, p. 46). Furthermore, this step will establish the communication between the interface and navigation in the hypermedia application.

- **Implementation**: In this stage, the hypermedia application will be implemented according to the user requirements and needs.

Table 6.19 illustrates that the design stage is very important for development of two key aspects of the website: navigational design and abstract interface design.

6.6.2   Implementation Methodology

Sampson et al. (2001) describe a methodology, which provides for the development of effective websites for counseling and career services. This methodology is very useful as it "can be used to consider opportunities for enhancing the design and use of the site" (Sampson et al. 2001) and it incorporates organizational aspects of implementation.

This model is divided into seven sequential stages, each of which must be completed before the next stage can be started. The stages are as follows:

- **Program evaluation**: This stage provides the foundation for the implementation process, helping to "ensures that the website is used for the right reasons with the right clients" (Sampson et al. 2001). The step has several sub-steps: evaluate the current resources and services; establish a committee; prepare an implementation plan; and seek stakeholder support.
- **Website development**: This stage helps the designer to make sure that the "website developed has the potential to effectively meet client and organization needs" (Sampson et al. 2001). The stage has several steps: develop and evaluate website contents and features and develop site documentation. In addition, this stage focuses on the development of website contents. Three questions need to be asked: "Whom does the website serve? What are the needs of users and what resources existing that would meet each of the identified needs?" (Sampson et al. 2001).
- **Website integration**: This stage involves the users to make sure that the website outcomes will meet their requirements. It begins with the "staff reviewing current

Table 6.20 Implementation methodology—extra stage—prepared by Tomayess Issa	Implementation methodology—extra stage
	Training Staff: from implementation model. This phase provides necessary training to the staff about the new system

needs and current resources and services" (Sampson et al. 2001), and then determining how the website will be used in delivering services and how it will operate according to user requirements.

- **Staff training**: Necessary training is given to the staff to incorporate the website with existing service delivery.
- **Trial use**: This stage requires the users to try out the website to see if it meets their needs. Moreover, continuing training is available in this stage, and observation and interview methods are used in order to determine if the website training is effective.
- **Operation**: This stage allows the user to operate and use the website.
- **Evaluation**: Evaluation and comments are collected from the users to ensure that the website services are running according to the user requirements. Therefore, the "results of the evaluation are used to indicate needed improvements in website design and use" (Sampson et al. 2001).

Finally, the feedback loops are indicated by the arrows and the staff responds to feedback as the implementation process continues. It was noted that this model is most suited to the development of websites for counseling and career services. However, it also has a wider application. This method includes a stage, which is essential to the system development process, which is training staff (see Table 6.20).

6.7 Summary of Information Systems Development Methodologies, Methodologies for Developing Websites, and Marketing Methodologies

New challenges have been imposed since the growth of use of the Internet as a global means of delivering information, selling goods, and entertainment. These new challenges suggest the need to develop a new methodology for developing websites which meet users' requirements and needs in order to avoid potential client frustration, make the website enjoyable, effective, and efficient, and most importantly, to improve performance (Table 6.21).

In this section, the researcher will summarize the results from the earlier analysis of information systems development methodologies, methodologies for developing websites, and marketing methodologies. The purpose behind the analysis is to:

- Identify the strongest stages of each methodology.
- Identify how well the four key principles are addressed in each methodology.
- Identify the extra stages from website and marketing methodologies.

Table 6.21 Summarizes the key aspects of the methodologies discussed in earlier sections of this chapter

Stage	Participation rating	Methodologies	Principles			
			User participation	Usability	Iteration	Real interaction
Planning	3	Soft System Methodology	1	0	2	0
		Human Factor Methodology for Designing Websites (HFMDW)	1	2	1	0
		Relationship Management Methodology (RMM)	0	0	1	0
		The Web Site Design Method (WSDM)	1	2	0	0
		E-Marketing Plan	1	0	0	0
		The Market-Vantage (Internet Performance Marketing) Methodology	0	0	0	0
Analysis	2	Soft Systems Methodology (SSM)	3	0	2	0
		User-Centered Development Methodology (UCDM)	1	0	0	0
		ETHICS Methodology	3	0	0	0
		Human Factor Methodology for Designing Websites (HFMDW)	0	3	1	2
		The Web Site Design Method (WSDM)	1	2	0	2
		Task Analysis				

(continued)

Table 6.21 (continued)

Stage	Participation rating	Methodologies	Principles			
			User participation	Usability	Iteration	Real interaction
Design	3	Structured Systems Analysis and Design Methodology (SSADM)	1	0	1	0
		Soft Systems Methodology (SSM)	2	0	2	0
		User-Centered Development Methodology (UCDM)	1	3	1	0
		Relationship Management Methodology (RMM)	0	0	2	0
		The W3DT Design Methodology	0	0	0	0
		The Web Site Design Method (WSDM)	1	2	0	2
		Navigation				
		Prototyping				
Testing	3	User-Centered Development Methodology (UCDM)	1	3	1	0
		Human Factor Methodology for Designing Websites (HFMDW)	1	3	2	0
		The Advertures Company Methodology	0	0	2	0

(continued)

Table 6.21 (continued)

Stage	Participation rating	Methodologies	Principles			
			User participation	Usability	Iteration	Real interaction
Implementation	2	Information Development Methodology for the Web	1	0	0	0
		E-Marketing Plan	1	0	0	0
		The Market-Vantage (Internet Performance Marketing) Methodology	0	0	2	0
		Construction				
		Promotion				
		Staff Training				
Evaluation	3	User-Centered Development Methodology (UCDM)	2	3	1	0
		Human Factor Methodology for Designing Websites (HFMDW)	0	3	2	0
		E-Marketing Plan	0	0	0	3
		Measurement				
Maintenance	2	Human Factor Methodology for Designing Websites (HFMDW)	0	0	1	3
		The Market-Vantage Methodology	2	0	0	2
		EnSky's Unique Methodology	1	0	0	1

Participation rate is from 0 to 3. Zero represents no participation while 3 indicates maximum participation. Ratings of 1 and 2 are minimum and moderate participation respectively. The ratings are based on the Consultative and Representative approaches according to Mumford (1995).

Identifying the strongest stage for each methodology will help the researcher to define the framework for the new participative methodology for developing websites.

The researcher identified several stages from the development lifecycle, which are: (1) planning, (2) analysis, (3) design, (4) testing, (5) implementation, (6) evaluation, and (7) maintenance. These stages are considered the basic and essential requirements for the system development process, as via these stages the designer will develop a system (interface or website) which meets the users' requirements.

Additionally, under the tables summarizing stages in the methodologies the researcher added four extra rows: "user participation," "usability," "iteration," and "real interaction." These key principles were either not fully considered in some methodologies or totally ignored. These principles are identified as being funda- mental to the proposed system development process of a website for marketing purposes, producing an effective interface or website. Simultaneously, through these principles, the designer and user will develop the new system (interface or website) to meet the user requirements and needs in order to make the design system flexible and adjustable, and to limit user frustration when working with it. These principles are the main foundation for this research.

The first row is **user participation**." It was noticed that user participation is a very practical approach in the development process. With it, the users will perform some activities and tasks and "these activities may pertain either to the management of the ISD project or to the analysis, design, and implementation of the system itself" (Hartwick and Barki 2001 p. 21).

Furthermore, according to Hartwick and Barki (2001), four dimensions of user participation can be identified: *RESPONSIBILITY; USER-IS RELATIONSHIP; HANDS-ON ACTIVITY*, and the most important aspect, which is *COMMUNICATION ACTIVITY*. These dimensions can deliver the following information to the designer.

- **Responsibility**: "the performance of activities and assignment reflecting overall leadership or accountability for the project."
- **User-IS Relationship**: "the performance of development activities reflecting users' formal review, evaluation and approval of work done by the IS staff."
- **Hand-On Activity**: "the performance of specific physical design and implemen- tation tasks."
- **Communication Activity**: "activities involving formal and informal exchange of facts, needs, opinions, visions, and concerns regarding the project among the users and between user and other project stakeholders" (Hartwick and Barki 2001 p. 22).

Therefore, the designer needs to work very closely with these dimensions in order to gain the basic information from the user about the system requirements and to identify the problems of the system. Furthermore, "user objectives, assumptions, strategies, actions, errors, problems, attitudes, etc., should surface so they can be explicitly considered in the system design and implementation processes" (Hartwick and Barki 2001 p. 22).

In addition, communication between the designers and users is an important aspect, which helps to identify the problems and to develop various solutions for the system by using different negotiation approaches and placing more emphasis on listening to users' needs and desires. For example, joint application development (JAD) workshops are "facilitated by a session leader trained in group dynamic techniques, where users and developers work together to plan and design a new system" (Hartwick and Barki 2001, p. 22).

The second row is "**usability**." This term is very important in the system development process as usability involves "an assortment of support for needs such as ease of use, ease of learning, error protection, graceful error recovery, and efficiency of performance" (Carroll 2002 p. 193). Usability will be emphasized in this research as it is considered very important especially in a methodology for developing websites.

The third row is "**iteration**." This term is very important in the system development process, as it can occur in each stage to ensure that the website is meeting the user requirements and company outcomes. This will enable the designers to build up the new website and make sure that the project will be tested repeatedly until it meets user requirements.

The fourth row is "**real interaction**." This term is very important in developing a website as it occurs in the maintenance and evaluation stages to ensure that user requirements are being met, by tracking use of the website by real users to achieve their specific objectives.

Finally, for the new participative framework for developing websites, a column will be added called "**participation rating**" which will help the researcher to identify the level of need for user participation in each stage. The participation rating will be from 0 to 3, indicating zero participation to maximum participation. The 1 and 2 ratings are minimum and moderate participation, respectively.

The researcher earlier reviewed the Mumford (1995) classification of user participation approaches in the system development process. In this research, the researcher will be using only the first two approaches: the consultative approach and the representative approach. Both of these approaches are very appropriate in all the stages in order to secure the agreement between users and designers at the beginning and to identify the key aspects, such as system objectives, problems, and the creating of various solutions to the system. The consensus approach will not be adopted in this research as it "does not always emerge easily and conflicts which result from different interests within a department may have to be resolved first" (Mumford 1995, pp. 18–19).

Extra stages were added from various methodologies for developing websites, mainly focusing on identifying user types, navigation, promotion, and prototyping. In addition, the researcher included more stages from marketing methodologies mainly focusing on promotion, prototyping, budget, return on investment (ROI), and measurement.

The requirements of a new participative methodology for developing websites include:

• Participation at all stages (different participation rates).

- Provision of detailed contents acquisition and maintenance requirements.
- Provision for detailed design of presentation.
- Provision of usability evaluation (at various stages).
- Provision of regular maintenance.

6.8 New Participative Methodology for Marketing Websites (NPMMW)

The New Participative Methodology for Marketing Websites (NPMMW) is developed from various existing models of system development and methodologies including lifecycle models, information systems development methodologies, methodologies for developing websites, marketing methodologies, and additional detailed techniques (see Figs. 6.1 and 6.2).

There are various comparisons with respect to the stages between methodologies for developing information systems, websites, or marketing strategies; however, integrating stages from information systems methodologies into a website with marketing methodologies is very valuable to improve websites that are more operative and effectual. User participation should be included in these methodologies to ensure

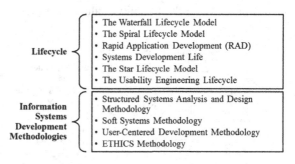

Fig. 6.1 Academic methodologies for development of Websites—prepared by the authors

Lifecycle
- The Waterfall Lifecycle Model
- The Spiral Lifecycle Model
- Rapid Application Development (RAD)
- Systems Development Life
- The Star Lifecycle Model
- The Usability Engineering Lifecycle

Information Systems Development Methodologies
- Structured Systems Analysis and Design Methodology
- Soft Systems Methodology
- User-Centered Development Methodology
- ETHICS Methodology

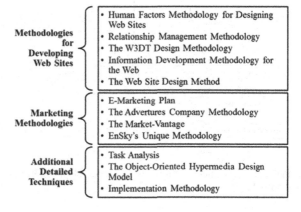

Fig. 6.2 Academic and commercial methodologies for development of Websites—prepared by the authors

Methodologies for Developing Web Sites
- Human Factors Methodology for Designing Web Sites
- Relationship Management Methodology
- The W3DT Design Methodology
- Information Development Methodology for the Web
- The Web Site Design Method

Marketing Methodologies
- E-Marketing Plan
- The Advertures Company Methodology
- The Market-Vantage
- EnSky's Unique Methodology

Additional Detailed Techniques
- Task Analysis
- The Object-Oriented Hypermedia Design Model
- Implementation Methodology

that transaction processes, tracking, maintenance, and updating of the website meet the users' requirements.

Each methodology was reviewed to determine two elements: (1) the stages needed for the system development process and (2) the utilization of four key principles (user participation, usability, iteration, and real interaction (i.e., the monitoring of user interaction with a prototype site). These principles were chosen to address the main deficits identified in existing website development methodologies and to produce a new methodology, which will assist in the development of websites with high usability.

The major stages of the New Participative Methodology for Marketing Websites (NPMMW) are presented in Fig. 6.3. Table 6.22 shows the issues, tools, and techniques for each stage and step, which need to be carried out by the designer in order to achieve a user-friendly website to prevent user frustration when she/he deals with this interface. The major stages of the methodology may be described as follows:

Usability Evaluation (SA0): This stage is located at the center of the new methodology, as, before the process moves on to another stage, it is necessary to evaluate the results from the previous stage, which is known as "formative evaluation." *Usability Evaluation—Measurement (SE0.1):* This step is an ongoing evaluation of the website to ensure that it achieves its intended purposes.

Functionality Testing (SA1): This stage is also located at the center of the new methodology (with the usability evaluation) to test the results from the previous

Fig. 6.3 New participative methodology for developing websites from the marketing perspective—Prepared by Tomayess Issa

Table 6.22 Issues, tools, and techniques for the new participative methodology—Prepared by Tomayess Issa

Stage (and step)	Issues, tools, and techniques
Usability evaluation	• Formative usability evaluation by expert- and user-based
Measurement	• *Ongoing evaluation*
Functionality testing	• Functionality testing by expert- and user-based

Planning	• Define the objectives
	  – User requirements
	  – User analysis
	  – Cost-benefit analysis
	  – Alternatives and constraints
	  – What is your product?
	  – Who are the buyers?
	  – Who are your competitors?
	  – Where should it be located?
	  – How to promote your website?

Analysis	• To add, improve, and correct the initial website requirements
Task analysis	• *Define user types, their work, goals, and activities*
Design	To define: • What the website is? • How the website will work to achieve the purpose behind using this website? • User involvement in decision making • Future users

Usability goals	• *User usability—Web design should be*
	  – *Efficient*
	  – *Effective*
	  – *Safe*
	  – *Utility*
	  – *Easy to learn*
	  – *Easy to remember*
	  – *Easy to use*
	  – *Easy to evaluate*

HCI goals	– *Usable* – *Practical* – *Visible* – *Job satisfaction* – *Extra techniques, text style, fonts, layout, graphics, and color*
Navigation	– *Site, layout, link, navigational structure for the hypermedia application*
Prototyping	– *High fidelity* – *Low fidelity*
Implementation	• Implementing the website using software
Construction	– *Technical application (i.e., HTML, Dreamweaver; Cold Fusion, and ASP)*
Training staff	– *Necessary training*

(continued)

Table 6.22 (continued)

Stage (and step)	Issues, tools, and techniques
Promotion	– *Press releases* – *Link building and banner ad campaigns* – *Paid search engine* – *Directory listing campaigns to promote the website* – *Traditional marketing (i.e., newspaper, radio, and TV)*
Maintenance	• Update changes and the corrector of errors on the website
Real interaction + feedback	– *Log file* – *Forms, survey, discussion forum, contact form, and telephone number*
Project review	– *Checklists*

stage before moving to another stage. Expert-based and user-based evaluations will test the website to ensure that it functions effectively from the technical perspective.

Planning (SA2): This stage allows designers and users to address various project-scoping issues: (1) the requirements for developing a website; (2) the nature of the product and the buyers; (3) the firm's competitors; and (4) the location of the site and how to promote the website. In addition, this stage involves developing a detailed schedule of activities required in order to carry out the development of the website in an efficient and effective manner.

Analysis (SA3): In this stage, users, analysts, and designers expand their findings in enough detail to indicate exactly what will and will not be built into the website design, and to add, improve, and correct the initial website requirements if they are not meeting the users' needs and wishes. Analysis—*Task Analysis (SE3.1):* This step will define the purpose of developing the website, the type of users, the type of work users will do with the website, users' goals, and their activities.

Design (SA4): The design stage will utilize the requirement specification from the previous stage to define: (1) what the website is; (2) how the website will work; (3) user involvement in decision making; (4) future users; and (5) usability requirements. *Design—Usability Goals (SE4.1):* This step will allow users (end-users and client–customer users), analysts, and designers (internal and external) to confirm that the website design is efficient, effective, safe, useful, easy to learn, easy to remember, easy to use and to evaluate, practical, and visible, and that it provides job satisfaction. *2 Design—HCI (SE4.2):* This step will allow users (end-users and client–customer users), analysts, and designers (internal and external) to identify that the website design is practical. There are many specific issues that need to be taken into consideration when designing website pages, such as text style, fonts, layout, graphics, and color. *Design—Navigation (SE4.3):* This step will define the specific navigation paths through the website among the entities to establish the communication between the interface and navigation in the hypermedia application. *Design—Prototyping (SE4.4):* This step is essential in the website design process, to allow users and management to interact with a prototype of the new website, to suggest changes, and to gain some experience in using it. This step allows the management to reduce costs and increase quality through early testing.

Implementation (SA5): This stage involves the technical implementation of the website design. It allows users to use the new product and to check whether it meets their requirements. *Implementation—Construction (SE5.1):* This step involves the technical implementation of the website design. *Implementation—raining Staff (SE5.2):* This step will give the necessary training to the staff about the new website. *Implementation—Promotion (SE5.3):* This step will use various tools such as press releases, link building and banner ad campaigns, paid search engines, directory listing campaigns, and traditional marketing methods (e.g., newspapers, radio, and TV) to promote the website.

Maintenance (SA6): This stage involves ongoing maintenance of the website, including updating changes and the correction of errors on the website. *Maintenance—Real Interaction and Feedback Tools (SE6.1):* During the maintenance stage, real interaction needs to be tracked by using the server log file. This information is very useful to the designers for improving and enhancing the structure and the functionality of the website in order to encourage more users to visit it. In addition, feedback tools should be available on the website to enable the users to contact the website owner for information or personal communication and to provide feedback about the website. For example, forms, surveys, discussion forum, contact form, telephone number, and a prize should be available on the website to encourage the users to provide feedback about the website. The first author recommends that, in order to prevent spam, the organization's e-mail address should not be made available on the website. *Maintenance—Project Review (SE6.2):* This step should be available to ensure that the website is working toward the project goals. This means that, after putting the website online, the designers need to check the website after one week to evaluate whether the website construction and structure are working according to the users' needs and requirements. One example of a tool that can be used for the project review is a checklist for the goals and objectives, usability, and technical requirements.

User Participation (SA7): This aspect is a very important concept in the methodology, as the main purpose is to allow user participation in the website development process in order to gain more information about the problems and alternative solutions from the users and to familiarize them with the system before it is released. For each stage, there is a rating (from 0 to 3), which indicates the extent of user participation in the development process.

Iteration (SA8): This occurs between each stage and step in the New Participative Methodology for Marketing Websites, to check that the website does indeed meet users' (end-users' and client–customer users') requirements and company objectives before moving to another stage.

Content Management Systems (CMS) (SA9): This aspect is relevant to the usability evaluation, functionality testing, planning, design, implementation, and maintenance stages in the New Participative Methodology for Marketing Websites. This tool will allow the users to manage the web contents by allowing them to add, edit, remove, and submit information by using various templates and workflows without needing any previous knowledge of the website editing tools.

6.9 Conclusion

This chapter has outlined the basic concepts behind methodologies including life-cycle models, IS development methodologies, methodologies with explicit human factors aspects, website methodologies, marketing methodologies, and additional detailed techniques such as task analysis and detailed website design and implementation. The main focus has been on defining users' requirements and needs, planning, analysis, design, testing, implementation, evaluation, and maintenance. These stages are very useful in any methodology, as, via them, the designer will make sure that the system is running according to the needs of users and the client organizations. In addition, four key principles (user participation, usability, iteration, and real interaction) were identified as fundamental aspects to develop systems in an effective manner. The four key principles are the main foundation for this research.

Having reviewed the stages from a wide range of methodologies, the chapter concludes with an introduction to the New Participative Methodology for Developing Websites from the marketing perspective combining the most effective aspects of the methodologies.

References

Advertures (2004) Process methodology. http://www.advertures.cz/alt/index_en.php?cat=company&sub=methodology. Accessed 27 Oct 2021

Avison D, Fitzgerald G (1993) Information systems development: methodologies. Alfred Waller Ltd., Publishers, UK, Techniques and Tools

Avison DE, Wood-Harper AT (1990) Multiview: an exploration in information systems development. Alfred Waller Ltd,UK

Bichler M, Nusser S, Wien W (1996) Modular design of complex web-applications with W3DT. In: 5th International workshops on enabling technologies: infrastructure for collaborative enterprises (WET ICE'96), Standord, CA. IEEE Proceedings of WET ICE '96, pp 328–333

Boyer MA (1999) Step 1: satisfy the consumer. Supermark Bus 54(4):112

Carroll JM (2002) Human-computer interaction in the new millennium. Addison-Wesley, New York

Checkland P (1981) Systems Thinking. Systems Practice. John Wiley & Sons Ltd., USA

Checkland P (2000) The emergent properties of SSM in use: a symposium by reflective practitioners. Syst Practice Action Res 13 (6)

Checkland P (1988) Information systems and systems thinking: time to unite? Int J Inf Manage 8(4):239–248

Checkland P, Scholes J (1990) Soft systems methodology in action. John Wiley & Sons Ltd., England

Checkland P, Scholes J (2003) Soft systems methodology in action. John Wiley & Sons, LTD, England

Cropley DH, Cook SC (1999) Systems methodology for real-time information systems. University of South Australia. http://www.unisa.edu.au/sec/pubs/99papers/222-P192.PDF. Accessed 27 Oct 2021

Darlington K (2005) Effective website development. Pearson Education Limited 2005,

De Troyer O (1998) Designing well-structured websites: lessons to be learned from database schema methodology. In: Tok Wang Ling SR, Mong Li Lee (eds) Conceptual modeling–ER '98: 17th International conference on conceptual modeling, Singapore, 16–19 Nov 1998. Springer-Verlag GmbH, Singapore, pp 51–64

De Troyer O, Leune C (1998) WSDM: a user centered design method for Web sites. In: Proceedings of the 7th international world wide web conference computer networks and ISDN systems. Elsevier, Vrijdag, pp 85–94

December J (1996) An information development methodology for the World Wide Web. Technical Communication 43(4):369

December J (2003) Developing information content for the World Wide Web. http://www.december.com/web/develop/overview.html. Accessed 27 Oct 2021

Dix A, Finlay J, Abowd G, Beale R (1998) Human-computer interaction, 2nd edn. Pearson Education Limited, USA

Doll WJ, Torkzadeh G (1989) A discrepancy model of end-user computing involvement. Manage Sci 35(10):1151

EnSky (1997) EnSky's unique methodology. http://www.ensky.com/company/process/methodology.php. Accessed 27 Oct 2021

FFIEC IT Examination Handbook (2005) Systems Development Life Cycle. Development and Acquisition, vol 9 November 2005.USA

Flynn DJ (1992) Information systems requirements: determination and analysis. McGraw-Hill, UK

Hartwick J, Barki H (2001) Communications as a dimension of user participation. IEEE Trans Prof Commun 44(1):21–36

Hirschheim R, Iivari J, Klein H (1998) A comparison of five alternative approaches to information systems development. http://www.bauer.uh.edu/parks/fis/sad5.htm. Accessed 27 Oct 2021

Hix D, Hartson HR (1993) Developing user interfaces: ensuring usability through product & process. John Wiley & Sons, USA

Iivari J, Hirschheim R, Klein HK (2001) A dynamic framework for classifying information systems development methodologies and approaches. J Manag Inf Syst 17(3):179–218

Instone K (2004) Usability engineering for the web. W3C J. http://www.w3j.com/5/s3.instone.html. Accessed 27 Oct 2021

Isakowitz T, Stohr E, Balasubramanian P (1995) RMM: a methodology for structured hypermedia design. Commun ACM 38(8):34–44

Isakowitz T, Stohr E, Balasubramanian P (2000) RMM: a methodology for structured hypermedia design. http://www.dgp.toronto.edu/~fanis/courses/hypermedia/rmm.html. Accessed 27 Oct 2021

Issa T (2008) Development and evaluation of a methodology for developing websites. PhD Thesis, Curtin University, Western Australia. http://espace.library.curtin.edu.au/R/MTS5B8S4X3B7SBAD5RHCGECEH2FLI5DB94FCFCEALV7UT55BFM-00465?func=results-jump-full&set_entry=000060&set_number=002569&base=GEN01-ERA02

Jayaratna N (1994) Understanding and evaluating methodologies -NIMSAD- a systemic framework. McGraw-Hill International, UK

Laux L (1998) Designing web pages and applications for people with disabilities. In: Chris Forsythe EGJR (ed) Human factors and web development. Lawrence Erlbaum Associates, Mahwah, NJ, pp 87–95

Lazar J (2006) Web usability: a user-centered design approach. Pearson Education Inc., USA

Market-Vantage (2003) Internet marketing methodology. http://www.market-vantage.com/about/methodology.htm. Accessed 27 Oct 2021

Mayhew DJ (1998) Human factors and the web. In: Chris Forsythe EGJR (ed) Human factors and web development. Lawrence Erlbaum Associates, Mahwah, NJ, pp 1–13

Mayhew DJ (1999) Business: strategic development of the usability engineering function. Interactions 6(5):27–34

McCracken DD, Wolfe RJ (2004) User-centered website development: a human-computer interaction approach. Pearson Education Inc., New Jersey, USA

Mumford E (1995) Effective systems design and requirements analysis. Macmillan Press Ltd., Great Britain

Mumford E (1996) Systems design: ethical tools for ethical change. Macmillan Press Ltd., Great Britain

Nielsen J (1992) The usability engineering life cycle. http://www-2.cs.cmu.edu/~jdh/courses/MOD/Neilesn-lifecycle.pdf. Accessed 27 Oct 2021

Olle TW, Hagelstein J, Macdonald IG, Rolland C, Sol HG, Assche FJMV, Verrijn-Stuart AA (1988) Information systems methodologies "a framework for understanding." Addison-Wesley Publishing Company, USA

Peters L (1988) Advanced structured systems analysis and design Prentice Hall. New Jersey, USA

Preece J, Rogers Y, Sharp H (2002) Interaction design: beyond human-computer interaction. John Wiley & Sons, USA

Rosenhead J, Mingers J (2002) Rational analysis for a problematic world revisited. John Wiley & Sons, Ltd, England

Sampson JP, Carr DL, Panke J, Arkin S, Minvielle M, Vernick SH (2001) An implementation model for web site design and use in counseling and career services. The Florida State University Tallahassee. http://www.career.fsu.edu/documents/implementation/Implementing%20Web%20Sites.ppt. Accessed 27 Oct 2021

Schwabe D, Rossi G (1995) The object-oriented hypermedia design model. Commun ACM 38(8):45–46

Strauss J, El-Ansary A, Frost R (2003) E-Marketing. Pearson Prentice Hall. www.nd.edu/~mkt384/mark461/powerpoints3/chapter3F.ppt. Accessed 27 Oct 2021

Strauss J, El-Ansary A, Frost R (2006) E-Marketing, 4th edn. Pearson Prentice Hall, Upper Saddle River

Vora P (1998) Human factors methodology for designing web sites. In: Chris Forsythe EGJR (ed) Human factors and web development. Lawrence Erlbaum Associates, Mahwah, NJ, pp 153–172

Chapter 7
New Participative Methodology for Sustainable Design (NPMSD)

Abstract Information and communications technology (ICT) use is increasing worldwide since ICT has become a significant mechanism for researching, searching, communication, entertainment, shopping, information, and more. However, the recycling of ICT products and the energy consumption of ICT is becoming a major problem for users and organizations locally and globally. Therefore, a solution is required to address this issue as a matter of urgency for the sake of the current and future generations. This chapter introduces and examines a New Participative Methodology for Sustainable Design (NPMSD). This design was assessed via an online survey conducted in Australia, Brazil, China, Germany, India, Norway, Singapore, South Korea, Sweden, UK, and USA. The survey outcomes confirmed the feasibility and value of the sustainable design step, and the online survey participants confirmed that through education and awareness, designers would learn more about sustainability and sustainable design.

7.1 Introduction

This chapter introduces and discusses and examines the NPMSD—a new methodology for sustainable design that will help to safeguard our planet. This methodology will assist designers to develop new smart technology and portable devices with sustainability features. Worldwide, the issues of recycling and energy consumption are causing major environmental problems by depleting natural resources, increasing the carbon footprint, and causing diseases and air pollution. Therefore, designers, academics, researchers, and individuals in general must understand their responsibility for our planet since there is no plan B for Earth. To tackle this problem, it is essential to raise designers' and HCI experts' awareness of their moral responsibility to create sustainable designs for a sustainable future. Designers of systems should take into account recycling, raw material supply, and energy consumption. This chapter is organized as follows: introduction, New Participative Methodology for Sustainable Design, and conclusion.

7.2   New Participative Methodology for Sustainable Design

In Chap. 1, based on the literature review, the initial sustainable step identified six factors regarding sustainability, namely design, safety, manufacture and energy, recycling, efficiency, and social needs. It is essential to consider these factors when developing a sustainable design for new smart technology and portable devices. Hence, designers and HCI experts should include these factors in their agenda to ensure that a good sustainable design will "eventually include criteria for the creation of a healthy environment and energy efficiency" (Stelzer 2006, p. 4).

Further studies (Atz et al. 2021; Comm and Mathaisel 2015; Dornfeld 2014; Kumar et al. 2021; Lo and Liao 2021; Melles et al. 2015; Pichlak and Szromek 2021; Russell-Smith et al. 2015; Wang et al. 2015a, b) confirm that integrating sustainability and corporate social responsibility (CSR) in any business strategy including the design process will enhance business reputation and preserve resources. Currently, sustainability and sustainable design are becoming the buzz words for users and organizations, as their inclusion and implementation in business strategies will be highly advantageous in terms of cost reduction, resources preservation, compliance with legislation, enhanced brand reputation, maintaining happier customers and stakeholders, attracting capital investment, and capitalizing on new opportunities (Weybrecht 2010). Kendall and Kendall (2010) indicated that sustainability will assist businesses, education, stakeholders, individuals, and society in general, while Jamwal et al. (2021) stated that the development of new technologies, i.e., industry 4.0, has assisted organizations to achieve sustainability due to a solid infrastructure.

Today, the world population exceeds 9.9 billion, and by 2050, it will be more than 10 billion (IISD 2020). This increase will influence the availability of housing, food, transportation, waste, economic and social issues, employment, the environment, and unsustainable development activities. Today, there is an urgent call for sustainable development in all areas including new smart technology and portable devices. Hence, to tackle these problems, IT and HCI experts should provide some solutions especially in design, manufacture, energy, waste management, and recycling by integrating and adopting sustainability in their design strategy especially for new smart technology and portable devices. It is a matter of urgency that designers and HCI experts be made fully aware of their moral responsibility to ensure sustainable development for a sustainable future.

A recent study by Internet World Stats (2021) found that 4.4 billion people are active Internet users, while Facebook users number 2.6 billion, representing a yearly increase of 27% and 10%, respectively. Table 7.1 shows the total number of active Internet users and mobile connections in various regions. These numbers are increasing daily, with a subsequent increase in the consumption of raw materials and the need for recycling.

As evident, the issue of sustainability does not concern only the environment, but extends to social and economic issues. Using an appropriate methodology and smart technology for designing new smart technology and portable devices will improve energy efficiency and reduce environmental impacts. Currently, increased usage of

Table 7.1 Digital usage by region (Internet World Stats 2021)

Region	Total population [Billion]	Active internet users [Billion]	Active Facebook users [Billion]
Asia	4,327,333,821	2,772,013,116	1,159,867,200
Africa	1,373,486	634,863	268,519,100
North America	370,322,393	347,916,627	264,153,100
South America	434,260,151	342,151,422	312,911,700
Europe	835,817,917	728,321,919	503,961,301
Middle East	265,587,661	201,014,130	147,065,800
Oceania And South Pacific	43,473,756	29,284,688	23,837,140

technology is becoming a pressing issue worldwide since technology has a huge impact on the environment as it consumes enormous amounts of raw materials and energy, produces greenhouse gases, and generates electronic waste that harms both the planet and mankind, causing serious diseases and death (Al Amri and Almaiah 2021; Issa et al. 2020; Maia 2021; Sharma et al. 2021; Shaw et al. 2015).

Individuals and organizations should understand that there is not another Earth to provide us with the resources and raw materials essential for our survival. Consequently, designers, users, and organizations should be mindful of the impact of their operations on the environment and take measures to become sustainable by integrating sustainability and sustainable design in their methodologies and strategies to reduce energy consumption and waste production and keep in mind the importance of recycling.

Further, it is essential for designers, users, and organizations to reorient their methodologies and strategies toward sustainable design and sustainability considering the environment problems that the world is currently facing. Finally, it is important for users, organizations, HCI experts, and designers to understand the impacts of their operations on the natural environment, particularly in regard to the use of technology. Therefore, it is imperative to take initiatives to address such problems via innovative and creative sustainable solutions achieved by educating users, organizations, HCI experts, designers, as well as top management about the importance of sustainable design and sustainability methodologies and strategies, which will increase technology performance and efficiency and reduce carbon emissions as well.

This book attempts to address these issues by introducing a new sustainable model for the design of smart technology and portable devices that can be applied now and in future. Therefore, the proposed NPMSD meets current needs without compromising the needs of future generations, since this NPMSD methodology offers various benefits in electric power reduction consumption of IT hardware and reduces CO2 emissions.

Currently, Bitcoin uses more energy than Amazon, Google, Microsoft, Facebook, and Apple combined. Bitcoin uses 121.05 TWh (terawatt-hours) per year, more than what is consumed by United Arab Emirates, the Netherlands, and Pakistan (Jackman 2021). Additionally, Lux Research found that Amazon, Microsoft, Google, Facebook, and Apple set a goal of becoming carbon neutral by 2030, this achievement needs 40 GW (GigaWatt) of renewable energy, or 20 GW of nuclear or 30 GW of mixed power sources in the next nine years. Currently, Microsoft has set the goal of becoming carbon negative. Microsoft established a $1 billion investment fund to achieve this target by 2030 (Conca 2021). In addition, Jordans (2021) indicates that the goal of carbon emission reduction is not limited to IT companies. Both developed and developing countries should take action to achieve this goal by 2030. According to a U.N. report, developed and developing countries that are signatories to the Paris climate accord until July 30, 2021, found that carbon emissions rising nearly 16% by 2030, compared with 2010 levels. Therefore, scientists worldwide are urging countries to take immediate measures to sharply curb carbon emission soon and add no more to atmosphere than can be absorbed by 2040; otherwise, the global temperature will rise by 1.5 °C (2.7 °F) by 2100. The UN data indicates that power consumption by large countries, organizations, and users in general is increasing at an alarming rate. Hence, a new methodology should be designed and applied to prevent or mitigate the undesirable outcomes related to manufacturing and energy consumption, by tackling the issues of design, recycling, safety, efficiency, and social impacts. These factors are taken into consideration in the NPMSD.

The NPMSD is driven by the New Participative Methodology for Marketing Websites (NPMMW) (Issa 2008). The NPMMW was developed from various existing models of system development and methodologies including lifecycle models, information systems development methodologies, methodologies for developing websites, marketing methodologies, and additional detailed techniques. The NPMMW is divided into ten stages, namely usability evaluation, functionality testing, planning, analysis, design implementation, maintenance, user participation, iteration, and content management systems. NPMMW is a contingency methodology as it allows users and designers to select the techniques, which best meet the requirements of the website, since each website from the marketing perspective has a different objective. To meet these objectives, the development of the website requires particular experience and skills. The NPMSD will apply the same principle to its various stages and includes a new step in the design process for the purpose of ensuring sustainability. This sustainability step addresses the issues of design, manufacture and energy, recycling, safety efficiency, and social impact (see Fig. 7.1).

The design factor is intended to facilitate upgrades and recycling and the addition of new software; most importantly, it ensures compliance with environmental standards and rules.

Regarding the manufacture and energy consumption factor, the new smart technology should tackle the energy issue by using less energy and raw materials and producing less waste and toxins. Moreover, the new smart technology should be powered by solar energy.

Fig. 7.1 Sustainable step—factors (prepared by Tomayess Issa)

In terms of the recycle factor, designers and HCI experts should use recycled, recyclable, and renewable materials to prevent the depletion of natural resources.

The safety factor aims to mitigate several negative outcomes of technology usage including carbon footprint, climate global warming, diseases, and air pollution. Therefore, the new smart technology design should consider these issues, especially in terms of recycling.

For the efficiency factor, designers and HCI experts should develop new smart technology and devices with long life, less packaging, and with convenient portability.

Finally, regarding the social factor, it is desirable to shift the mode of consumption from personal ownership of products to the provision of services, clean emissions, successful production cycles, and good ethical principles.

If the aforementioned factors are taken into consideration in new smart technology and portable devices design, resources for the next generation will be conserved, and our planet will be safeguarded from toxic pollutants and diseases. These factors are considered in the sustainability step in the design stage, which comprises usability goals, HCI, navigation, and prototyping.

The major stages and steps of the New Participative Methodology for Sustainable Design are presented in Fig. 7.2. Table 7.2 shows the issues, tools, and techniques for the activities need to be carried out by the designer when developing a sustainable design. These are described below:

Usability Evaluation (SA0): this stage is essential, as before proceeding to another stage, it is necessary to evaluate the outcomes of the previous stage; this is known as "formative evaluation."

- *Usability Evaluation—Measurement (SE0.1)*: this step is an ongoing evaluation of the new device to ensure that it will achieve its intended purpose(s).

Fig. 7.2 New participative methodology for sustainable design—prepared by Tomayess Issa

Functionality Testing (SA1): this stage is also essential to the new methodology (with the usability evaluation) as it tests the results from the previous stage before moving to another stage. Expert-based and user-based evaluations will test the new device to ensure that it functions effectively from the technical perspective.

Planning (SA2): this stage allows designers and users to address various project-scoping issues: (1) the requirements for developing a new device, (2) the nature of the product and the buyers, (3) the firm's competitors. In addition, this stage involves developing a detailed schedule of activities required in order to carry out the development of the new devices in an efficient and effective manner.

Analysis (SA3): in this stage, users, analysts, and designers expand their findings in enough detail to indicate exactly what will and will not be built into the device design and to add, improve, and correct the initial device requirements if they are not meeting the users' needs and wishes.

- *Analysis—Task Analysis (SE3.1)*: this step will define the purpose of developing the device, the type of users, the type of work users will match with the device users' goals, and their activities.

Design (SA4): the design stage will utilize the requirement specifications from the previous stage to determine: (1) what the device is; (2) how the device will work; (3) user involvement in decision making; (4) future users; (5) usability requirements.

- *Design—Usability Goals (SE4.1)*: this step will allow users (end-users and client–customer users), analysts, and designers (internal and external) to confirm that the device design is efficient, effective, safe, useful, easy to learn, easy to remember, easy to use and to evaluate, practical, and visible, and that it provides job satisfaction.

Table 7.2 Stages, steps and issues, tools and techniques for the new participative methodology for sustainable design—prepared by Tomayess Issa

Stage (and step)	Issues, tools, and techniques
Usability evaluation	• Formative usability evaluation by expert and user based
Measurement	• Ongoing evaluation
Functionality testing	• Functionality testing by expert- and user-based
Planning	• Define the objectives • User requirements • User analysis • Cost-benefits analysis • Alternatives and constraints • What is your product? • Who are the buyers? • Who are your competitors? • Where should it be located? • How to promote your Smart Technology or portable device?
Analysis	• To add, improve, and correct the initial smart technology or portable device requirements
Task analysis	• Define user types, their work, goals, and activities
Design	To define: • What the smart technology or portable device is • How the smart technology or portable device will work to achieve the purpose of the new technology • User involvement in decision making • Future users

Usability goals	• User usability—smart technology or portable device design should be
	  – Efficient
	  – Effective
	  – Safe
	  – Utility
	  – Easy to learn
	  – Easy to remember
	  – Easy to use
	  – Easy to evaluate

HCI goals	– Usable – Practical – Visible – Job satisfaction – Extra techniques, text style, fonts, layout, graphics, and color

(continued)

Table 7.2 (continued)

Stage (and step)	Issues, tools, and techniques
Sustainable	• Design – Easy to upgrade – Easy to add new software – Easy to recycle – Complies with environmental standards and regulations • Manufacture and Energy – Uses less energy – Uses solar energy – Uses less raw materials – Produces less waste and toxins • Recycle – Uses recycled materials – Uses recyclable materials – Uses renewable materials – Recycle – Uses recycled materials – Uses recyclable materials – Uses renewable materials • Safety – Reduces carbon footprint – Reduces global warming – Reduces diseases and even death of humans – Reduces air pollution – Reduces consumption and waste of resources • Efficiency – Have long life – Have less packaging – Have portability efficiency • Social – Shifts the mode of consumption from personal ownership of products to provision of services – Has clean emissions – Has successful production cycles – Has good ethical principles
Navigation	– Site in portal device, Layout, Link, Navigational Structure for the hypermedia Application in the portal device
Prototyping	– High-fidelity – Low-fidelity
Implementation	Implementing the smart technology or portable device
Construction	– Technical applications
Training users	– Necessary Training

(continued)

Table 7.2 (continued)

Stage (and step)	Issues, tools, and techniques
Promotion	– Press releases – Link building and banner ad campaigns – Paid search engine – Directory listing campaigns to promote the smart technology or portable device – Traditional marketing (i.e., newspapers, radio, and TV) – Digital marketing (i.e., internet and social networking tools)
Maintenance	• Update changes and the correct of errors
Project review	– Checklists

- *Design—HCI (SE4.2)*: this step will allow users (end-users and client–customer users), analysts, and designers (internal and external) to identify that the device design is practical. There are many specific issues that need to be taken into consideration when designing a device, such as text style, fonts, layout, graphics, and color.
- *Design—Sustainable (SE4.3)*: this step will allow designers to consider the factors necessary for developing new smart technology and portable devices comprising sustainability factors.
- *Design—Navigation (SE4.4)*: this step will define the specific navigation paths through the device among the entities to establish the communication between the interface and navigation in the hypermedia application.
- *Design—Prototyping (SE4.5)*: this step is essential in the device design process as it allows users and management to interact with a prototype, to suggest changes, and to gain some experience in using it. This step allows the management to reduce costs and increase final quality through early testing.

Implementation (SA5): this stage involves the technical implementation of the device design. It allows users to use the new product and to check whether it meets their requirements.

- *Implementation—Construction (SE5.1)*: this step involves the technical implementation of the new smart technology and portable device design.
- *Implementation—Training Users (SE5.2)*: this step will give the necessary training to users of the new smart technology.
- *Implementation—Promotion (SE5.3)*: this step will use various communication tools such as press releases, link building and banner ad campaigns, paid search engines, directory listing campaigns, and traditional marketing methods (e.g., newspapers, radio, and TV) and digital marketing methods (i.e., Internet and social networking, i.e., Facebook, Twitter, Instagram, and Snapchat) to promote the new smart technology.

Maintenance (SA6): this stage involves ongoing maintenance of the device.

- *Maintenance—Project Review (SE6.1)*: this step ensures that the device is working toward achieving the project goals. This means that, after the device becomes "alive", the designers need to check the device after one week to determine whether the device construction and structure are working according to the users' needs and requirements. One example of a tool that can be used for the project review is a checklist for the goals and objectives, usability, and technical requirements.

User Participation (SA7): this aspect is a very important concept in the methodology, as the main purpose is to allow prospective users to participate in the device development process in order to identify and problems and suggest alternative solutions and to become familiar with the device before it is released. For each stage, there is a rating (from 0 to 3), which indicates the extent of user participation in the development process.

Iteration (SA8): this occurs between each stage and step in the NPMSD to check that the device does indeed meet users' requirements and company objectives before moving to the next stage.

To assess the feasibility and value of the sustainable step and its factors, an online survey was conducted in Australia, Brazil, China, Germany, India, Norway, Singapore, South Korea, Sweden, UK, and USA. The online survey questionnaire was based on the literature review and consisted of two parts: background and sustainable design. The survey was distributed to the participants through the Qualtrics website (www.qualtrics.com). Qualtrics is an online survey tool that has a reliable reputation for developing and summarizing survey results; it allows researchers to collect and analyze data obtained online (Boas and Hidalgo 2013). The online survey data was analyzed using SPSS version 27.

Table 7.3 shows the number and percentage of online survey participants in terms of gender, age, and qualifications. The survey response rate was 100% and 51% are female. The majority of respondents (18.65%) were aged between 25 and 30 years, while 33.72% (a majority) have a bachelor degree.

Table 7.4 shows the technology used by the participants. The online survey results indicated that 41.47% of online survey participants are spending up to five hours per day on the computer for professional work and study with 51.26% on the Internet. Furthermore, 69% spend less than an hour on email per day, while 62.7% spend less than an hour daily on social networking.

Furthermore, the online survey found that 14.42% of the participants are using the Internet to access their email, while 12.57% use it for shopping online and 11.13% for banking online (see Table 7.5).

Furthermore, the online survey identified the devices used to access the Internet. 25.73% of respondents are using a smartphone, and 24.28% are using a laptop (Table 7.6).

Data showed that participants were first introduced to the concepts of sustainability and green information technology via Internet (27.69%), news media (22%), and school (15.69%) (see Table 7.7).

Table 7.3 Online survey statistics—prepared by Tomayess Issa

Number and percentage of online survey	
Questionnaires distributed	1705
Questionnaires returned	1705
Response rate	100%
Developed countries	1190
Developing countries	515
Gender	
Male respondents	835 (48.97%)
Female respondents	870 (51.03%)
Age	
17 years and under	0 (0%)
18–20	116 (6.8%)
21–24	176 (10.32%)
25–30	318 (18.65%)
31–35	288 (16.89%)
36–40	178 (10.44%)
41–45	131 (7.68%)
46–50	139 (8.15%)
51–55	128 (7.51%)
56–60	127 (7.45%)
61–65	104 (6.10%)
Over 65	0 (0%)
Qualifications	
Primary rducation	74 (4.34%)
Higher secondary/pre-university	285 (16.72%)
Professional certificate	117 (6.86%)
Diploma	132 (7.74%)
Advanced/higher/graduate diploma	111 (6.51%)
Bachelor's degree	575 (33.72%)
Postgraduate diploma	83 (4.87%)
Master's degree	225 (13.20%)
Ph.D	44 (2.58%)
Others	59 (3.46%)

Table 7.4 Technology use—online survey statistics—prepared by Tomayess Issa

Answer	Hours spent on the computer per day	Hours spent on the Internet per day	Hours spent on the email per day	Hours spent on the social networking per day
	Response %	Response %	Response %	Response %
Less than an hour	72 (4.22%)	88 (5.16%)	459 (26.92%)	550 (32.26%)
Up to five hours	707 (41.47%)	874 (51.26%)	12 (0.70%)	19 (1.11%)
Five to ten hours	665 (39%)	527 (30.91%)	2 (0.12%)	7 (0.41%)
Ten to twenty hours	221 (12.96%)	190 (11.14%)	1178 (69.09%)	1069 (62.7%)
Over twenty hours	40 (2.35%)	26 (1.52%)	54 (3.17%)	60 (3.52%)

Table 7.5 Internet usage—online survey statistics—prepared by Tomayess Issa

Answer	Response	%
Email	1618	14.42
Playing games	736	6.56
Studying	753	6.71
Working	1046	9.32
Shopping online	1411	12.57
Chatting	877	7.82
Researching hobbies	910	8.11
Banking online	1249	11.13
Buying goods or services	1178	10.50
Buying stocks or investing online	384	3.42
Researching travel information or making reservations	945	8.42
Others—please specify	115	1.02

The online survey examined the companies which were associated with online survey users' devices. Google and Apple are the leaders at 34.72% and 25.57%, respectively (see Table 7.8).

Participants were asked whether they read the sustainability report of the manufacturer before buying a device. The survey shows that 38.65% do not read the report; 33.26% responded "maybe", and 19.65% read the report. This outcome indicates that users should take more responsibility for their actions and be aware of their responsibility to the planet. This awareness needs to be increased via education and training (see Table 7.9).

Table 7.6 Device's usage—online survey statistics—prepared by Tomayess Issa

Answer	Response	%
PC	879	16.78
Desktop	721	13.76
Laptop	1272	24.28
Netbook	140	2.67
PDAs	36	0.69
Workstation	84	1.6
Tablet	733	13.99
Smartphone	1348	25.73
Others—please specify	26	0.5

Table 7.7 First introduced to the concepts of sustainability and green information technology—online survey statistics—prepared by Tomayess Issa

Answer	Response	%
School	519	15.69
Higher education	382	11.55
Internet	916	27.69
Books	241	7.29
Magazine	295	8.92
News media	730	22
Conferences	110	3.33
Others—please specify	115	3.48

Table 7.8 Online survey users' devices—prepared by Tomayess Issa

Answer	Response	%
Apple	436	25.57
Google	609	35.72
Dell	203	11.91
IBM	64	3.75
Others—please specify	393	23

Table 7.9 Online survey users reading the sustainability report of the company before buying a device—prepared by Tomayess Issa

Answer	Response	%
Yes	335	19.65
No	659	38.65
Maybe	567	33.26
Not at all	144	8.45

Table 7.10 Online survey users changing their device—prepared by Tomayess Issa

Answer	Response	%
Every six months	35	2.05
Every 12 months	169	9.91
Every 18 months	212	12.43
Every 24 months	429	25.16
Every 30 months	111	6.51
Every 36 months	265	15.54
Every 42 months	241	14.13
Other—please specify	243	14.25

Table 7.11 Online survey users "Why do you change your device"—prepared by Tomayess Issa

Answer	Response	%
Size	859	25.89
Speed	235	7.08
Functionality	1099	33.12
Keeping up with technology	919	27.70
Others—please specify	206	6.21

In addition, the online survey users change their device after 24 to 42 months with percentages ranging from 25 to 14%, respectively (see Table 7.10).

When asked their reason for changing devices, 33% of participants indicated that new devices offered more functions; 27.7% wanted to keep up with technology, and for 25.89% of participants, size was the determining factor (see Table 7.11).

Additionally, the online survey sought to determine the participants' attitudes to their moral responsibilities toward the planet by asking whether they believed that changing their devices frequently will cause damage to our planet. 41.94% participants were aware that changing devices frequently would cause damage to our planet, while 41.76% responded "Maybe" (see Table 7.12).

Table 7.12 Online survey users "changing device frequently will cause damage to our planet" prepared by Tomayess Issa

Answer	Response	%
Yes	715	41.94
No	246	14.43
Maybe	712	41.76
Not at all	32	1.88

Table 7.13 Online survey users: "can we change the mindset of designers and users regarding sustainability"—prepared by Tomayess Issa

Answer	Response	%
Training	684	10.62
Education	1282	19.90
Awareness	1209	18.77
Workshop	466	7.23
Internet	1056	16.39
T.V	763	11.85
Social networking	916	14.22
Others—please specify	65	1.01

Participants were asked to recommend ways to change the mindset of designers and users toward sustainability. The survey concluded that via education and awareness (19.90% and 18.77%, respectively) designers and users could change their mindset and attitude (see Table 7.13).

A total of *1705 valid cases* were processed for the subsequent factor analysis. The analysis was conducted separately for two groups with 23 for sustainable design and 37 questions for the Green IT advantages and disadvantages of sustainability, respectively. The first part for the group sustainable design consists of six groups/aspects based on users' level of awareness of sustainable design adopted from Stelzer (2006, p. 4). Those aspects are design [4 questions]; safety [5 questions]; manufacture and energy [4 questions]; recycling [3 questions]; efficiency [3 questions]; and social factors [4 questions]. For the advantages, the aspects are financial [5 questions] brand and reputation [9 questions], human resources and shareholders [6 questions], and environmental [4 questions]. For the disadvantages, the items related to failure [8 questions] and cost [5 questions]. To further examine the online survey data, the researchers adopted principal axis factoring for factor extraction, and oblique rotation (rather than orthogonal rotation) was applied using the Promax method (Costello and Osborne 2005; Field 2016, 2013; Hair et al. 2009). Cronbach's alpha, Kaiser–Meyer–Olkin, and Bartlett's test of sphericity were applied to measure sampling adequacy. Table 7.14 shows the statistical results obtained by these tests. Cronbach's alpha for all 23 variables from sustainable design awareness is 0.945, while the alpha for advantages and disadvantages are 0.938 and 0.932, respectively, indicating that all items in the scale had excellent internal consistency (Amirrudin et al. 2021; Gliem and Gliem 2003; Vaske et al. 2017). A Kaiser–Meyer–Olkin (KMO) measure of sampling adequacy ranging from 0.954 to 0.928 indicates that a marvelous sample size has been obtained for the analysis (Hill 2012; Shrestha 2021); hence, the current KMO results are appropriate and acceptable for this study. The Bartlett's test of sphericity is highly significant, indicating that the items of the scale are sufficiently correlated for factors to be found (Shrestha 2021; Tobias and Carlson 1969; Williams et al. 2010). The results shown in Table 7.14 indicate the validity and rationality of this step.

Table 7.14 Alpha, KMO, and BT: sustainable design awareness, advantages, and disadvantages—prepared by Tomayess Issa

Theme	Cronbach's Alpha	KMO sampling adequacy	Bartlett's test of sphericity
Sustainable design awareness, advantages, and disadvantages—all countries			
Sample Size 1705			
Sustainable design awareness	0.945 [Excellent]	0.954 (Marvelous)	$\chi^2 = 23{,}127.988\ df = 253\ p < 0.000$
Sustainable design advantages	0.938 [Excellent]	0.951 (Marvelous)	$\chi^2 = 22{,}040.715;\ df = 276\ p < 0.000$
Sustainable design disadvantages	0.932 [Excellent]	0.934 (Marvelous)	$\chi^2 = 151{,}418.585;\ df = 78\ p < 0.000$
Sustainable design awareness, advantages, and disadvantages—developed countries			
Sample Size 1190			
Sustainable design awareness	0.944 [Excellent]	0.950 (Marvelous)	$\chi^2 = 16{,}635.249\ df = 253\ p < 0.000$
Sustainable design advantages	0.935 [Excellent]	0.945 (Marvelous)	$\chi^2 = 15{,}697.824\ df = 276\ p < 0.000$
Sustainable design disadvantages	0.928 [Excellent]	0.928 (Marvelous)	$\chi^2 = 10{,}348.771\ df = 78\ p < 0.000$
Sustainable design awareness, advantages, and disadvantages—developing countries			
Sample size 515			
Sustainable design awareness	0.945 [Excellent]	0.949 (Marvelous)	$\chi^2 = 6765.584\ df = 253\ p < 0.000$
Sustainable design advantages	0.930 [Excellent]	0.934 (Marvelous)	$\chi^2 = 5944.265\ df = 276\ p < 0.000$
Sustainable design disadvantages	0.939 [Excellent]	0.935 (Marvelous)	$\chi^2 = 5089.142\ df = 78\ p < 0.000$

The researchers used principal components analysis to estimate the factor-loading matrix for the factor analysis model as well as the standard correlation matrix. The eigenvalues were assessed to determine the number of factors accounting for the correlations among the variables. For the sustainable design awareness section, Table 7.14 shows the total variance values for sustainable design awareness, advantages, and disadvantages of all countries, developed and developing with a range of 51.370% to 72.969% of the variation. Table 7.14 shows that the variance is divided between two and three components to be extracted for these variables. Williams et al. (2010) stated that the explained cumulative proportion of variance is usually 50–60%, which is considered acceptable for variance explained. The amount of variance explained by each of these components is presented in Table 7.15 (after rotation).

In order to measure the regression coefficients (i.e., slopes), the researchers carried out factor loadings. The factor loadings of most of the items were adequately high, and the one with the cleanest fact structured is considered to be important (Costello

Table 7.15 Total variance for sustainable design awareness, advantages, and disadvantages section—prepared by Tomayess Issa

Component	Initial eigenvalues			Extraction sums of squared loadings			Rotation sums of squared loadings		
	Total	% of variance	Cumulative %	Total	% of variance	Cumulative %	Total	% of variance	Cumulative %
Total variance explained—all countries *sustainable design awareness*									
1	10.629	46.212	46.212	10.629	46.212	46.212	8.428	36.645	36.645
2	1.507	6.553	52.764	1.507	6.553	52.764	3.707	16.119	52.764
Total variance explained—developed countries *sustainable design awareness*									
1	10.583	46.012	46.012	10.583	46.012	46.012	8.347	36.293	36.293
2	1.552	6.747	52.758	1.552	6.747	52.758	3.787	16.465	52.758
Total variance explained—developing countries *sustainable design awareness*									
1	10.637	46.247	46.247	10.637	46.247	46.247	8.751	38.046	38.046
2	1.491	6.482	52.729	1.491	6.482	52.729	3.377	14.683	52.729
Total variance explained—all countries—advantages									
1	9.943	41.427	41.427	9.943	41.427	41.427	6.736	28.065	28.065
2	2.386	9.943	51.370	2.386	9.943	51.370	5.593	23.305	51.370
Total variance explained—developed countries—advantages									
1	9.749	40.620	40.620	9.749	40.620	40.620	5.220	21.749	21.749
2	2.509	10.453	51.073	2.509	10.453	51.073	5.135	21.396	43.145
3	1.279	5.331	56.404	1.279	5.331	56.404	3.182	13.259	56.404
Total variance explained—developing countries—advantages									
1	9.298	38.742	38.742	9.298	38.742	38.742	6.112	25.467	25.467
2	2.290	9.541	48.284	2.290	9.541	48.284	3.947	16.445	41.912
3	1.220	5.083	53.366	1.220	5.083	53.366	2.749	11.455	53.366

(continued)

Table 7.15 (continued)

Component	Initial eigenvalues			Extraction sums of squared loadings			Rotation sums of squared loadings		
	Total	% of variance	Cumulative %	Total	% of variance	Cumulative %	Total	% of variance	Cumulative %
Total variance explained—all countries—*disadvantages*									
1	7.220	55.536	55.536	7.220	55.536	55.536	3.303	25.408	25.408
2	1.428	10.982	66.518	1.428	10.982	66.518	3.252	25.017	50.424
3	0.839	6.451	72.969	0.839	6.451	72.969	2.931	22.544	72.969
Total variance explained—developed countries—*disadvantages*									
1	7.041	54.164	54.164	7.041	54.164	54.164	3.201	24.622	24.622
2	1.453	11.177	65.341	1.453	11.177	65.341	3.100	23.849	48.471
3	0.885	6.809	72.150	0.885	6.809	72.150	3.078	23.680	72.150
Total variance explained—developing countries—*disadvantages*									
1	7.561	58.159	58.159	7.561	58.159	58.159	4.097	31.512	31.512
2	1.372	10.550	68.709	1.372	10.550	68.709	3.465	26.652	58.164
3	0.807	6.205	74.915	0.807	6.205	74.915	2.178	16.750	74.915

Extraction method: principal component analysis

and Osborne 2005). Factor items with a loading less than 0.5 were excluded 0.5 based on the rule of thumb suggested by Stevens (1992) for a sample size above 100. In addition, Rose et al. (2011) suggested that the acceptable factor loading based on sample size between 200–249 is 0.40, and is 0.30 for a sample size above 1000. However, in this study, the factor loading was selected as 0.6 to capture relevant statements to develop a new factor (see Table 7.16). Table 7.16 shows the group pattern matrix for the sustainable design awareness section.

Table 7.16 shows the new factors for sustainable design awareness, advantages, and disadvantages. For the sustainable design awareness, two new identical factors, namely safe design and sustainable smart design, were derived from data for all countries, both developed and developing. The study participants indicated that safe design aims to reduce climate global warming, carbon footprint, air pollution, and waste of resources, but the most important aspect is the reduction of waste and toxins on our planet. While the majority of participants indicated that the sustainable smart design of new devices should be easy to upgrade by adding new software to the current devices, in which case, less materials will be required to create new devices, users can keep their devices for long periods, and most importantly, these devices will be up to date to meet the current and new software and application needs.

Following the analysis of the survey data, a slight change was made to the sustainable step in the design stage. The design sub-step was renamed sustainable smart design sub-step, the safe sub-step was renamed safe design sub-step, and these changes are highlighted in green in Table 7.17.

Based on the study outcomes, Fig. 7.3 presents the additional sustainable step, under the design stage. The changes are presented in black font.

Furthermore, the study examined the sustainable design advantages and disadvantages for all countries, developed and developing. Based on the data analysis, a new common factor was generated, namely environmental value among all countries, developed and developing. This factor aims to reduce pollution, emissions, carbon footprint, and health hazard. This factor must be considered by organizations, designers, and user, since this will help us to achieve the UN sustainability goals now and in future (Pouget 2021).

Additionally, Table 7.18 the new factors to generated in regard to the advantages of sustainable design: productivity, CSR, meet stakeholders needs, and sustainable fiscal, these advantages provided a clear picture that integrating sustainable design in organizations strategy will make the organizations more unique, and competitive and supporting their country sustainability strategy locally and globally.

On the other hand, Table 7.18 indicated that incorporation sustainable design concept in the organization's strategy can cause some disadvantages, namely increase security and risk and complain environmental policy, regulations, and cost. These disadvantages can cause a huge rejection by the organizations, users, designers, and HCI experts to disseminate the sustainable design idea in the strategy. To minimize this impact, awareness and training are very essential to spread the good news about sustainable design advantages, as is known any new project there is a risk behind it, but with good planning and preparation these disadvantages can be minimized.

Table 7.16 Rotated component matrix—sustainable design awareness, advantages, and disadvantages—prepared by Tomayess Issa

All countries—sustainable design awareness

Rotated component matrix[a]

	Component		
	1	2	
Reduce climate global warming	0.797		Safe design
Reduce carbon footprint	0.781	0.144	
Reduce consumption and waste of resources	0.775	0.186	
Reduce air pollution	0.741	0.129	
Produce less waste and toxins	0.730	0.282	
Sustain environmental standards and rules	0.720	0.223	
Use renewable materials	0.697	0.337	
Use recyclable materials	0.697	0.303	
Have clean emissions	0.658	0.367	
Use recycled materials	0.657	0.314	
Are easy to recycle	0.626	0.245	
Use less energy	0.621	0.320	
Have good ethical principles	0.615	0.333	
Use less raw materials	0.604	0.284	
Reduce diseases and even death of humans	0.576	0.196	
Have less packaging	0.560	0.407	
Have successful production cycles	0.529	0.484	
Use solar energy	0.526	0.328	
Are easy to add new software		0.826	Sustainable smart design
Are easy to upgrade		0.809	
Have portability efficiency	0.376	0.590	
Have long life	0.425	0.537	
Shifting the mode of consumption from personal ownership of products to provision of services	0.298	0.431	
Extraction method: principal component analysis Rotation method: Varimax with Kaiser normalization			
a. Rotation converged in three iterations			

(continued)

Table 7.16 (continued)

Developed countries—sustainable design awareness

Rotated component matrix[a]

	Component		
	1	2	
Reduce consumption and waste of resources	0.794	0.159	Safe design
Reduce carbon footprint	0.787	0.132	
Reduce climate global warming	0.786	0.120	
Sustain environmental standards and rules	0.748	0.188	
Produce less waste and toxins	0.738	0.294	
Use renewable materials	0.720	0.326	
Use recyclable materials	0.705	0.303	
Reduce air pollution	0.695	0.179	
Are easy to recycle	0.695	0.194	
Use recycled materials	0.671	0.319	
Use less raw materials	0.637	0.277	
Have good ethical principles	0.621	0.327	
Use less energy	0.619	0.305	
Have clean emissions	0.615	0.425	
Have less packaging	0.599	0.380	
Reduce diseases and even death of humans	0.482	0.267	
Use solar energy	0.480	0.346	
Are easy to add new software	0.128	0.783	Sustainable smart design
Are easy to upgrade		0.773	
Have portability efficiency	0.288	0.649	
Have long life	0.431	0.549	
Have successful production cycles	0.491	0.511	
Sustainable design produces devices that: Shifting the mode of consumption from personal ownership of products to provision of services	0.184	0.504	
Extraction method: principal component analysis Rotation method: Varimax with Kaiser normalization			
a. Rotation converged in three iterations			

(continued)

Table 7.16 (continued)

Developing countries—sustainable design awareness

Rotated component matrix[a]

	Component		
	1	2	
Reduce climate global warming	0.777		Safe design
Reduce air pollution	0.762	0.111	
Reduce carbon footprint	0.760	0.161	
Reduce consumption and waste of resources	0.745	0.215	
Produce less waste and toxins	0.740	0.182	
Use recyclable materials	0.730	0.195	
Have clean emissions	0.726	0.252	
Use renewable materials	0.681	0.291	
Use recycled materials	0.678	0.205	
Sustain environmental standards and rules	0.678	0.258	
Use less energy	0.670	0.279	
Reduce diseases and even death of humans	0.663	0.193	
Use solar energy	0.581	0.323	
Have successful production cycles	0.580	0.456	
Have good ethical principles	0.568	0.388	
Use less raw materials	0.568	0.241	
Have less packaging	0.548	0.404	
Are easy to recycle	0.516	0.291	
Have long life	0.463	0.448	
Shifting the mode of consumption from personal ownership of products to provision of services	0.421	0.415	
Are easy to add new software		0.847	Sustainable smart design
Are easy to upgrade		0.843	
Have portability efficiency	0.514	0.518	
Extraction method: principal component analysis Rotation method: Varimax with Kaiser normalization			
a. Rotation converged in three iterations			

(continued)

Table 7.16 (continued)

All countries—sustainable design advantages

Rotated component matrix[a]

	Component		
	1	2	
Attract quality employees	0.742	0.165	Productivity and superiority
Increase productivity	0.732	0.107	
Satisfy customer needs	0.719	0.158	
Meet stakeholder expectations	0.715		
Improve community investments	0.684	0.267	
Attract new opportunities	0.681	0.236	
Improve human rights	0.636	0.238	
Create new jobs	0.629	0.230	
Improve social responsibility investing	0.621	0.386	
Reduce risk management	0.618	0.156	
Increase triple bottom line—people, planet, and profit	0.574	0.392	
Increase cost-effectiveness	0.550	0.288	
Differentiate businesses	0.538	0.239	
Increase efficiency	0.533	0.333	
Enhance reputation	0.511	0.345	
Reduce pollution	0.185	0.836	Environmental value
Reduce emissions	0.170	0.835	
Reduce carbon footprint	0.195	0.812	
Reduce energy and water usage	0.163	0.742	
Reduce health hazards	0.297	0.690	
Reduce consumption of raw materials	0.196	0.688	
Reduce paper usage	0.236	0.639	
Increase green strategy	0.393	0.585	
Improve corporate social responsibility	0.437	0.534	
Extraction method: principal component analysis Rotation method: Varimax with Kaiser normalization			
a. Rotation converged in three iterations			

(continued)

Table 7.16 (continued)

Developed countries—sustainable design advantages

Rotated component matrix[a]

	Component			
	1	2	3	
Reduce emissions	0.835	0.224		Environmental value
Reduce pollution	0.833	0.251		
Reduce carbon footprint	0.806	0.278		
Reduce energy and water usage	0.752	0.105	0.205	
Reduce consumption of raw materials	0.745		0.251	
Reduce health hazards	0.664	0.321	0.185	
Reduce paper usage	0.623	0.145	0.269	
Improve social responsibility investing	0.307	0.713	0.118	CSR
Improve community investments	0.174	0.666	0.258	
Attract new opportunities	0.179	0.658	0.287	
Improve human rights	0.180	0.606	0.245	
Differentiate businesses	0.195	0.598		
Attract quality employees	0.112	0.583	0.463	
Create new jobs	0.168	0.567	0.326	
Enhance reputation	0.324	0.543		
Increase green strategy	0.524	0.531		
Increase triple bottom line—people, planet, and profit	0.319	0.530	0.273	
Improve corporate social responsibility	0.501	0.511		
Meet stakeholder expectations		0.505	0.487	
Reduce risk management		0.500	0.410	
Increase efficiency	0.348	0.102	0.766	Financial success
Increase cost-effectiveness	0.281	0.174	0.740	
Increase productivity		0.386	0.687	
Satisfy customer needs	0.142	0.466	0.559	

Extraction method: principal component analysis
Rotation method: Varimax with Kaiser normalization

a. Rotation converged in ten iterations

(continued)

Table 7.16 (continued)

Developing countries—sustainable design advantages

Rotated component matrix[a]

	Component			
	1	2	3	
Satisfy customer needs	0.736		0.130	Meet stakeholders need
Meet stakeholder expectations	0.708		0.126	
Attract quality employees	0.702		0.273	
Attract new opportunities	0.686	0.252		
Improve community investments	0.679	0.264	0.197	
Increase productivity	0.678		0.222	
Improve human rights	0.650	0.186	0.125	
Enhance reputation	0.631	0.214	0.206	
Differentiate businesses	0.618	0.182	0.129	
Improve social responsibility investing	0.596	0.389	0.131	
Create new jobs	0.592	0.272		
Increase triple bottom line—People, Planet, and Profit	0.525	0.421	0.195	
Reduce risk management	0.453		0.407	
Improve corporate social responsibility	0.432	0.374	0.351	
Reduce pollution	0.155	0.816	0.177	Environmental value
Reduce emissions	0.170	0.810	0.171	
Reduce carbon footprint	0.209	0.779	0.160	
Reduce health hazards	0.178	0.667	0.232	
Increase green strategy	0.348	0.584	0.222	
Reduce consumption of raw materials	0.169	0.220	0.718	Sustainable fiscal
Reduce energy and water usage		0.441	0.687	
Reduce paper usage		0.391	0.676	
Increase efficiency	0.412	0.104	0.500	
Increase cost-effectiveness	0.392	0.101	0.499	

Extraction method: principal component analysis
Rotation method: Varimax with Kaiser normalization

a. Rotation converged in seven iterations

All countries—sustainable design disadvantages

(continued)

Table 7.16 (continued)

Rotated component matrix[a]

	Component			
	1	2	3	
Increase marketing failure by perceived environmental irresponsibility	0.796	0.346	0.278	Failure risk
Increase scandals by perceived environmental irresponsibility	0.775	0.340	0.289	
Increase supply chain crises due to suppliers' environmental problems	0.757	0.313	0.340	
Increase security and systems failures caused by environmental problems	0.740	0.369	0.181	
Increase litigation and compliance breaches (including environmental actions and environmental compliance failures)	0.226	0.856	0.218	Fraud and complain
Increase fraud (including environmental actions and environmental compliance failures)	0.275	0.815	0.169	
Increase governance failure (including environmental actions and environmental compliance failures)	0.340	0.815	0.204	
Increase transaction failure due to environmental liabilities	0.398	0.735	0.221	
Increase number of new regulations including environmental		0.121	0.820	Increase environmental policy
Increase competition for and cost of raw materials	0.158	0.201	0.764	
Increase interest rates	0.346	0.209	0.665	
Inflate costs	0.393	0.162	0.631	
Increase insurance crises due to environmental disasters	0.490	0.239	0.598	
Extraction method: principal component analysis Rotation method: Varimax with Kaiser normalization				
a. Rotation converged in five iterations				

(continued)

The online survey outcomes indicated that all countries of developed and developing are encouraging sustainable design for the current technology, new smart technology, and portable devices, by asking designers and HCI experts to integrate and adopt sustainability and sustainable design concepts in their design process. The study participants need designers to preserve raw resources and materials for the

Table 7.16 (continued)

Developed countries—sustainable design disadvantages

Rotated component matrix[a]

	Component			
	1	2	3	
Increase litigation and compliance breaches (including environmental actions and environmental compliance failures)	0.843	0.227	0.224	Failure and complain
Increase governance failure (including environmental actions and environmental compliance failures)	0.817	0.334	0.189	
Increase fraud (including environmental actions and environmental compliance failures)	0.791	0.258	0.179	
Increase transaction failure due to environmental liabilities	0.756	0.340	0.246	
Increase marketing failure by perceived environmental irresponsibility	0.311	0.816	0.286	Marketing failure
Increase scandals by perceived environmental irresponsibility	0.302	0.792	0.302	
Increase supply chain crises due to suppliers' environmental problems	0.301	0.761	0.341	
Increase security and systems failures caused by environmental problems	0.394	0.694	0.179	
Increase number of new regulations including environmental	0.154		0.787	Increase environmental policy
Increase competition for and cost of raw materials	0.201	0.166	0.740	
Inflate costs	0.114	0.307	0.726	
Increase interest rates	0.219	0.272	0.708	
Increase insurance crises due to environmental disasters	0.256	0.444	0.620	
Extraction method: principal component analysis Rotation method: Varimax with Kaiser normalization				
a. Rotation converged in six iterations				

(continued)

future generations, known as the seventh generation. Furthermore, the study participants confirmed that sustainable design is the way to go in the future, and we need to make users more aware of the consequences for future generations regarding sustainability by raising awareness through education and training, since the biggest problem is that most people seem to want the latest products on the market.

Table 7.16   (continued)

Developing countries—sustainable design disadvantages

Rotated component matrix[a]

	Component			
	1	2	3	
Increase security and systems failures caused by environmental problems	0.797	0.366		Security failure
Increase scandals by perceived environmental irresponsibility	0.766	0.417	0.181	
Increase marketing failure by perceived environmental irresponsibility	0.758	0.429	0.199	
Increase supply chain crises due to suppliers' environmental problems	0.755	0.356	0.278	
Inflate costs	0.638	0.219	0.350	
Increase insurance crises due to environmental disasters	0.605	0.247	0.489	
Increase interest rates	0.598	0.181	0.441	
Increase litigation and compliance breaches (including environmental actions and environmental compliance failures)	0.246	0.868	0.183	Fraud and complain
Increase fraud (including environmental actions and environmental compliance failures)	0.306	0.846	0.126	
Increase governance failure (including environmental actions and environmental compliance failures)	0.357	0.813	0.205	
Increase transaction failure due to environmental liabilities	0.478	0.721	0.115	
Increase number of new regulations including environmental	0.191		0.857	Increase regulations and cost
Increase competition for and cost of raw materials	0.230	0.222	0.790	
Extraction method: principal component analysis Rotation method: Varimax with Kaiser normalization				
a. Rotation converged in five iterations				

Figure 7.4 depicts the new factors for sustainable design comprising awareness, advantages, and disadvantages which are generated from specific developed and developing countries. Finally, the survey outcomes confirmed the author's views on the sustainable design obtained by the new participative methodology.

Table 7.17 Sustainable step—design stage: before and after the current study

Before: sustainable step: issues, tools, and techniques	After: sustainable step: issues, tools, and techniques

• Design
  – Easy to upgrade
  – Easy to add new software
  – Easy to recycle
  – Complies with environmental standards and regulations
• Manufacture and energy
  – Uses less energy
  – Uses solar energy
  – Uses less raw materials
  – Produces less waste and toxins
  – Recycle
  – Uses recycled materials
  – Uses recyclable materials
  – Uses renewable materials
• Safety
  – Reduces carbon footprint
  – Reduces global warming
  – Reduces diseases and even death of humans
  – Reduces air pollution
  – Reduces consumption and waste of resources
• Efficiency
  – Has long life
  – Has less packaging
  – Has portability efficiency
• Social
  – Shifts the mode of consumption from personal ownership of products to provision of services
  – Has clean emissions
  – Has successful production cycles
  – Has good ethical principles
	• Sustainable Smart Design
	  – Easy to upgrade
	  – Easy to add new software
	  – Easy to recycle
	  – Complies with environmental standards and regulations
	• Manufacture and energy
	  – Uses less energy
	  – Uses solar energy
	  – Uses less raw materials
	  – Produces less waste and toxins
	  – Recycle
	  – Uses recycled materials
	  – Uses recyclable materials
	  – Uses renewable materials
	• Recycle
	  – Uses recycled materials
	  – Uses recyclable materials
	  – Uses renewable materials
	• Safe design
	  – Reduces carbon footprint
	  – Reduces climate global warming
	  – Reduces diseases and even death of humans
	  – Reduces air pollution
	  – Reduces consumption and waste of resources
	• Efficiency
	  – Have long life
	  – Have less packaging
	  – Have portability efficiency
	• Social
	  – Shifts the mode of consumption from personal ownership of products to provision of services
	  – Has clean emissions
	  – Has successful production cycles
	  – Has good ethical principles

7.3 Conclusion

This chapter is concerned with the development of a New Participative Methodology for Sustainable Design and establishing a sustainable design process which comprises design, manufacture and energy, recycling, safety efficiency, and social factors. This methodology was developed to raise designers' and users' awareness of sustainability and green information technology in terms of technology and portable

Fig. 7.3 Sustainable step in sustainable design—Prepared by Tomayess Issa

Table 7.18 New factors of sustainable design awareness, advantages, and disadvantages—prepared by Tomayess Issa

New factors	All countries	Developed	Developing
New factors—sustainable design awareness	Safe design Sustainable smart design	Safe design Sustainable smart design	Safe design Sustainable dmart design
New factors—sustainable design advantages	Productivity and superiority Environmental value	Environmental value CSR Financial success	Meet stakeholders need Environmental value Sustainable fiscal
New factors—sustainable design disadvantages	Failure risk Fraud and complain Increase environmental policy	Failure and complain Marketing failure Increase environmental policy	Security failure Fraud and complain Increase regulations and cost

devices design. The application of this methodology to the design of devices and new smart technology will reduce the harm done to our planet as a result of poor recycling and the over-consumption of energy and raw materials. Finally, we academics have a responsibility to increase our students' awareness and make them part of the solution not the problem, encouraging them to serve their countries and communities while being good stewards of the planet. In the future, additional research will be carried out to implement and assess the sustainable design step with large IT companies to ensure compliance with environmental standards and regulations for sustainable

Fig. 7.4 Summary of the new factors for sustainable design: awareness, advantages, and disadvantages

systems. Finally, our planet is suffering from our actions, and we need to address this issue before it is too late to save the planet, as there is NO Plan B for the Earth (Issa et al. 2020).

References

Al Amri M, Almaiah MA (2021) Sustainability model for predicting smart education technology adoption based on student perspectives. Int J Adv Soft Comput Appl 13(2)

Amirrudin M, Nasution K, Supahar S (2021) Effect of variability on Cronbach alpha reliability in research practice. Jurnal Matematika, Statistika Dan Komputasi 17(2):223–230

Atz U, Van Holt T, Douglas E, Whelan T (2021) The return on sustainability investment (ROSI): monetizing financial benefits of sustainability actions in companies. In: Sustainable consumption and production, vol II. Springer, pp 303–354

Boas TC, Hidalgo FD (2013) Fielding complex online surveys using Rapache and Qualtrics. Polit Methodol 20(2):21–26

Comm CL, Mathaisel DF (2015) Designing an engineering system for sustainability. Applied mechanics and materials. Trans Tech Publ, pp 474–478

Conca J (2021) Big tech companies look to clean energy to slake their enormous Thirst. https://www.forbes.com/sites/jamesconca/2021/05/24/big-tech-companies-look-to-clean-energy-to-slake-their-enormous-thirst/?sh=ba25b9356155. Accessed 21 Oct 2021

Costello A, Osborne J (2005) Best practices in exploratory factor analysis: four recommendations for getting the most from your analysis. Pract Assess Res Eval 10(7):1–9

Dornfeld DA (2014) Moving towards green and sustainable manufacturing. Int J Precis Eng Manuf Green Technol 1(1):63–66

Field A (2013) Discovering statistics using IBM SPSS statistics. SAGE

Field A (2016) An adventure in statistics: the reality enigma. Sage

Gliem JA, Gliem RR (2003) Calculating, interpreting, and reporting Cronbach's alpha reliability coefficient for Likert-type scales. In: Midwest research-to-practice conference in adult, continuing, and community education

Hair J, Black W, Babin B, Anderson R (2009) Multivariate data analysis. Prentice Hall Upper Saddle River, NJ

Hill BD (2012) The sequential Kaiser-Meyer-Olkin procedure as an alternative for determining the number of factors in common-factor analysis: a Monte Carlo simulation proquest. Umi Dissertation Publishing USA

IISD (2020) World population to reach 9.9 billion by 2050 https://sdg.iisd.org/news/world-popula tion-to-reach-9-9-billion-by-2050/. Accessed 21 Oct 2021

Internet World Stats (2021) Internet world stats https://www.internetworldstats.com/. Accessed 21 Oct 2021

Issa T (2008) Development and evaluation of a methodology for developing websites. PhD Thesis, Curtin University, Western Australia. http://espace.library.curtin.edu.au/R/MTS5B8S4X3B7 SBAD5RHCGECEH2FLI5DB94FCFCEALV7UT55BFM-00465?func=results-jump-full&set_ entry=000060&set_number=002569&base=GEN01-ERA02

Issa T, Issa T, Issa TB, Isaias P (2020) Sustainability awareness and green information technologies. Springer

Jackman J (2021) Bitcoin uses more energy than Amazon, Google, Microsoft, Facebook, and Apple combined. https://www.theecoexperts.co.uk/blog/bitcoin-uses-more-energy-than-amazon. Accessed 21 Oct 2021

Jamwal A, Agrawal R, Sharma M, Kumar V, Kumar S (2021) Developing a sustainability framework for industry 4.0. Procedia CIRP 98:430–435

Jordans F (2021) United Nations warns of hot and 'catastrophic' future. https://www.nationalobse rver.com/2021/09/18/news/united-nations-warns-hot-and-catastrophic-future. Accessed 21 Oct 2021

Kendall K, Kendall J (2010) Forms of government and systemic sustainability: a positive design approach to the design of information systems advances in appreciative. Inquiry 3:137–155

Kumar S, Sureka R, Lim WM, Kumar Mangla S, Goyal N (2021) What do we know about business strategy and environmental research? Insights from business strategy and the environment. Business Strategy and the Environment

Lo F-Y, Liao P-C (2021) Rethinking financial performance and corporate sustainability: perspectives on resources and strategies. Technol Forecast Soc Change 162:120346

Maia M (2021) A perspective on 21th design challenges facing the urgency of solutions for sustainability global problems the emerging of an education and design profession alignment about core principles. In: 6 CIDAG, 2021

Melles G, Kuys B, Kapoor A, Rajanayagam J, Thomas J, Mahalingam A (2015) Designing technology, services and systems for social impact in the developing world: strong sustainability required. In: Chakrabarti AE (ed) ICoRD'15–research into design across boundaries, vol 2. Springer, pp 89–97

Pichlak M, Szromek AR (2021) Eco-innovation, sustainability and business model innovation by open innovation dynamics. J Open Innov Technol Market Complex 7(2):149

Pouget R (2021) Call for 'decisive action now' to avoid climate catastrophe. https://news.un.org/ en/story/2021/09/1100382. Accessed 21 Oct 2021

Rose S, Hair N, Clark M (2011) Online customer experience: a review of the business-to-consumer online purchase context. Int J Manag Rev 13:24–39

Russell-Smith SV, Lepech MD, Fruchter R, Littman A (2015) Impact of progressive sustainable target value assessment on building design decisions. Build Environ 85:52–60

Sharma RR, Kaur T, Syan AS (2021) Sustainability and business environments. In: Sustainability marketing. Emerald Publishing Limited

Shaw G, Walters R, Kumar A, Sprigg A (2015) Sustainability in infrastructure asset management. In: Proceedings of the 7th world congress on engineering asset management (WCEAM 2012). Springer, pp 525–534

Shrestha N (2021) Factor analysis as a tool for survey analysis. Am J Appl Math Stat 9(1):4–11

Stelzer K (2006) Sustainability = good design. Les Ateliters De Lethique 2:1-15

Stevens J (1992) Applied multivariate statistics for the social sciences. Erlbaum, Hillsdale, NJ

Tobias S, Carlson JE (1969) Brief report: Bartlett's test of sphericity and chance findings in factor analysis. Multivar Behav Res 4(3):375–377

Vaske JJ, Beaman J, Sponarski CC (2017) Rethinking internal consistency in Cronbach's Alpha. Leis Sci 39(2):163–173. https://doi.org/10.1080/01490400.2015.1127189

Wang L, Kwok JS, Tsang DC, Poon C-S (2015a) Mixture design and treatment methods for recycling contaminated sediment. J Hazard Mater 283:623–632

Wang L, Tsang DC, Poon C-S (2015b) Green remediation and recycling of contaminated sediment by waste-incorporated stabilization/solidification. Chemosphere 122:257–264

Weybrecht G (2010) The sustainable MBA. In: The manager's guide to green business. Wiley, England, UK

Williams B, Onsman A, Brown T (2010) Exploratory factor analysis: a five-step guide for novices. J Emerg Primary Health Care (JEPHC) 8(3):1–13

Chapter 8
Innovative Technologies: Applications in the Present and Considerations for the Future

Abstract Information and communication technologies are evolving rapidly at an unprecedented rate, with repercussions on all areas of society. Developments in key innovative technologies are at the center of revolutionary advancements in education, business, health, and government. At the same time, technological progress poses diverse challenges that can compromise inclusivity and sustainability and increase security, ethical and privacy concerns. This chapter examines the key technological trends associated with social networks, artificial intelligence, Internet of Things and big data, and extended reality. It explores their most recent developments and applications and the challenges they pose across several sectors. It also examines the fundamental role that sustainability and the digital divide play in an era of galloping technological development.

8.1 Introduction

Information and communication technologies (ICTs) have become central to today's information-driven society. They include a variety of technologies with different benefits that greatly shape all sectors of society but have their own specific set of challenges. The present accepts the value of ICTs and reaps the rewards of their adoption, while the future questions and reflects on the problems and issues arising from their continued use.

A panoply of innovative technologies is revolutionizing society. Social networks, artificial intelligence (AI), Internet of Things (IoT) and big data, and extended reality (XR) are not novel technologies, but they were selected for this chapter due to the innovations that they continuously introduce into society. Social networks continue to attract a large number of Internet users (Peng et al. 2018) and are valuable for communication and community building (Rosabel et al. 2022) AI continues to give greater capabilities to computers, enabling them to perform tasks in a human-like manner across numerous fields (Chen et al. 2020). IoT's importance is notable in many fields (i.e., transportation, health, finance) (Nord et al. 2019), while the volume, velocity, and variety of big data enables information to be extracted that is valuable to businesses, government departments, and the like (Oussous et al. 2018). Finally,

XR's contribution is intrinsically connected with its enhancement of reality (Gong et al. 2021).

This chapter examines some of the most relevant and innovative technologies, although the list is not exhaustive. It investigates the roles that social networks, AI, IoT and big data, and XR play in various sectors and the challenges they pose. It begins by investigating the various contributions that social networks can make to education and business, as well as the problems they raise. It then examines the role of AI in education and health and addresses related issues and ethics. IoT and big data follow their value to smart cities and e-government is discussed, along with the difficulties they present. Next, the benefits offered by XR to the health and business sectors are examined, as well as its associated challenges. The final section reflects on the future of ICTs in terms of sustainability and the digital divide.

8.2 Social Networks in the Information Age

Social networks have undergone significant development and are attracting an increasing number of users. Social networks can be defined as networks that depict the various nodes and connections of a social structure. They comprise online social networks, existing in online environment and mobile social networks, which are based on mobile applications. Various other social networks have specific functions; among these are social news platforms or media-sharing applications (Peng et al. 2018). Social networks serve as valuable instruments in several sectors and numerous organizations, such as small and medium enterprises (Olvera-Lobo 2020) and non-profit organizations (Pífano et al. 2021). They are also used for advertising (Falcão and Isaías 2020) and for health (Alshaikh et al. 2014).

8.2.1 Implications for Business

Within online social networks, there are members who are considered to be "influencers." These members can be very valuable for businesses due to their word-of-mouth power and their status as role models. They can reach their contacts more efficiently, enabling them to disseminate information swiftly and widely; because they are seen as role models, other members are more likely to imitate them. The identification of these key users has become a central issue for business, so much so that the strategies for that identification have recently become a significant research topic (Klein et al. 2015). Moreover, social networks are important sources of information for brands. The analysis and mining of data from these platforms can offer valuable insights into their users' opinions about a specific product or service, mainly through the comments posted by users (Peng et al. 2018). Online social networks have an important role in the empowerment of clients. They are interactive platforms that allow users to generate content, search for information, and express their opinions about different products and brands. Internet users are sometimes called "digital evangelists" due to their influential role in social networks, which can cause a product to either succeed or fail. Also, because they offer suggestions regarding new products or services, they are often known as "prosumers" for their part in companies' creative process (Gonzalez et al. 2015). Regardless of their advantages, the successful deployment of social network sites in the business arena should follow specific guidelines. Moreover, it is crucial to use measurable criteria to assess the actual effects that online social networks have on revenue (Isaías et al. 2012).

The user engagement with web-based social networks has repercussions on their business relationships. It is believed that individuals with an online social network presence have more opportunities to connect and strengthen their ties with other professionals. Despite being hosted online, web-based social networks are facilitating offline relationships (Benson et al. 2014). As the most popular professional social network, LinkedIn has attracted much interest and has an impressive number of users worldwide. The use of LinkedIn is positively related to networking ability. Also, frequent users, rather than those with a high number of contacts, can expect to obtain career benefits from this social network (Davis et al. 2020). Social networks are excellent communication channels with unlimited audience reach and information dissemination. When examining the dynamics of event organization, for example, the important role that social networks play in event promotion is very evident. Organisers can use social networks as vehicles of information. In the case of music festivals, a significant amount of data can be disseminated through social networks (performers, schedules, etc.) to those attending or wishing to attend the event. Additionally, the engagement of people in social networks is potentially beneficial in terms of building the attendees' loyalty to the event and again in terms of marketing the event with personal statements (text, photographs, etc.) provided by the attendees (Hudson et al. 2015).

8.2.2 Social Networks in the Education Sector

Distance and mobile education are becoming pervasive with the assistance of various existing and emerging (Isaías 2018; Miranda et al. 2017) learning technologies (Isaias et al. 2017), including social networks (Shelomovska et al. 2017). Social networks can be used for both formal or informal education (Teoh et al. 2014) and across different levels of education, such as in secondary schools (Ng 2021) and in universities (Quimbita et al. 2020). Both teachers and students have expressed positive attitudes toward the use of social networks in higher education, regarding them as valuable tools for teaching and learning, and are willing to increase their use in educational settings (Shestak et al. 2021). The adoption of social networks by the higher education sector can strengthen students' engagement and improve their motivation to learn. In addition, even their learning performance can be improved, increasing their level of achievement (Hortigüela-Alcalá et al. 2019). The constraints imposed by the safety measures adopted during the COVID-19 pandemic lockdown periods has led to an increase in the use of ICTs. In the current environment, social networks enable a large number of users to be connected and are efficient instruments of communication that help to create meaningful online learning communities (Rosabel et al. 2022).

Social networks are being used in the education sector, but education is also being used in the online social networks arena. The increase in the number of children and teenagers who use social networks has led to the design of several educational packages that promote a more secure participation on these platforms (Vanderhoven et al. 2014). In the context of online learning, in particular within Massive Open Online Course (MOOC), the students use tools such as social networks to engage in collaborative learning and to build learning communities. Also, on these platforms, they engage with tagged content and use the existing tagging features to maintain or establish new conversations to enhance their learning (Cruz-Benito et al. 2017). Microlearning, which promotes the delivery of learning via brief learning units, is one of the fields where social networks can be very effective. Social networks connect users with a diversity of interests and different levels of knowledge, enabling collaboration. Moreover, social networks enable the swift and widespread exchange of content and empowers students to interact with each other based on the shared content (Giurgiu 2017). Social networks can be valuable instruments for students in the sense that they allow the creation of networks of users, and they provide features enabling the exchange of content. In addition, students do not need to be trained on how to use them, since they are very familiar with these platforms and use them frequently (Imran et al. 2016).

8.2.3   Social Networks Main Challenges

Despite their many advantages, the use of social networks presents several challenges (Persia and Auria 2017). The nature of social networks has given rise to two major issues: privacy and anonymity and the dissemination of misinformation and disinformation. In regard to privacy and anonymity, some online social networks impose a real-name policy which prevents their users from using an alias. This policy is intended as a strategy to improve content and service, to facilitate users' search for contacts, and to enable accountability. Despite the benefits that social networks often cite to justify this policy, there is a growing controversy associated with the use of the user's real identity. By requesting their users to register with their real identity, these platforms have access to information that jeopardizes privacy and online freedom (Peddinti et al. 2014). Anonymity can become problematic, for example, in the use of location-based services, where the geographical position of the social network users is used to provide services. Furthermore, besides potentially exposing customers who wish to remain anonymous, these services can include information that poses a threat to the privacy of the users' data (Buccafurri et al. 2021). The increasing number of social network users causes great volumes of varied information to be posted online, which increases the availability of datasets via the Internet and the uncertainty about whether that data is protected from de-anonymization. In light of this predicament, transparency is emerging as a new framework for information management. In addition to transparency, the right to be forgotten is vital in information management in the sense that it would allow users to delete previous data, when introducing new information (Kataoka et al. 2014).

Social networks, which enable information to "go viral," have the capacity to influence decision making. There are insufficient mechanisms to protect social network users against misinformation and its consequences (Bastick 2021). Misinformation can have serious repercussions on many levels. In some cases, it can influence voting decisions, cause instability in financial markets, impact the decision of parents regarding vaccination for their children, and create panic and spread incorrect information about disease outbreaks, as is currently occurring with the COVID-19 pandemic (Amoruso et al. 2020).

There is increasing interest in the dissemination of misinformation within social networks. Bastick (2021) conducted an experiment with 233 undergraduate students and concluded that even in cases where the students were exposed briefly (less than five minutes) to fake news their unconscious behavior was substantially modified. Social networks are among the most effective instruments for information dissemination. They are used not only as key communication channels among individuals for both personal and professional purposes, but they are also the main source of news for a substantial number of Internet users (Amoruso et al. 2020). The detection of fake news is often done via content, by verifying the accuracy of the news, with some websites specializing in this type of service. Nonetheless, this individual verification of news is a time-consuming and complex process, particularly when there is a high volume of content being shared swiftly and widely. By combining content with methods that analyze the credibility of the source, the author can be more productive (Sitaula et al. 2020). There are also methods for the detection of fake news that are style-based, in that the writing style is analyzed, and propagation-based, focusing on information offered by the dissemination of the news (Zhou et al. 2019).

8.3 AI and the Barrier Between Human and Machines

AI can be defined as a "system's ability to interpret external data correctly, to learn from such data, and to use those learnings to achieve specific goals and tasks through flexible adaptation" (Kaplan and Haenlein 2019, p. 15). With AI, computers become capable of performing human-like tasks. AI is used across several fields (Chen et al. 2020) including marketing (Huang and Rust 2021), industry (Ahmad et al. 2021), agriculture (Zhang et al. 2021), and entertainment (Lachman and Joffe 2021).

8.3.1 AI in Educational Settings

The possibilities offered by new formats of learning online, such as MOOCs, translate into an increase in the number of students, greater student participation, and consequently rising costs for institutions. In these scenarios, automated solutions seem to be a valuable instrument to address the challenges that emerge, such as the need to provide personalized support to large numbers of students. Teachers can be integrated into online and blended courses to assist with the delivery of content, to offer feedback, and to supervise the students (Popenici and Kerr 2017). Despite its benefits, AI is still far from being adopted widely in education. In higher education, risk aversion, resistance to innovation, and insufficient funding are some of the obstacles to the incorporation of novel technology. In addition, educators often require convincing of the true value of technology before adopting it. AI can become more pervasive if educators become more aware of its benefits and shortcomings, whether there is an increased multidisciplinarity between different types of educational experts and technology experts and whether AI applications can become more aligned with contemporary educational theories (Bates et al. 2020). AI can be used with big data to create smart learning environments where teachers can analyze students' progress, identify those who might be at risk of failing, and intervene in a timely manner. AI has the potential to improve both the teachers' performance, by improving their pedagogical practices, and students' learning experience, by empowering them to be more in control of their learning process. The value of the information provided by AI translates into actionable insight obtained from concrete data about the students' difficulties, performance, and progress (Yang et al. 2021). As an example, it is happening with learning analytics and the valuable insight it offers education (Quadir et al. 2020; Clark et al. 2020).

The use of AI in education is known as artificial intelligence education (AIEd). This term refers to the general application of AI in educational settings, namely through the deployment of various AI-based applications and systems already in place in many schools and higher education institutions (Sun et al. 2021). In education, AI has been associated with a series of benefits, evident, for example, at the level of personalization, grading, quality of teaching, and student learning experience. AI-based learning systems integrate several techniques for analysis, recommendation, and understanding, supported by a knowledge model, machine learning, and data mining. These systems have the ability to map the connections between the students' results and factors such as the learning resources and the teacher's approach. The information they provide enables teachers to customize their teaching methods and strategies to improve the students' learning outcomes and experience. Through automation, AI can assist teachers to grade students' exams and offer feedback on assignments (Chen et al. 2020). AI can be used to offer personalized content to the students, according to their profile and preferences, at the same time that it can assist teachers with course design. Moreover, one the fundamental benefit of AI is the fact that by performing tasks previously done by teachers, the AI system frees the educators from time-consuming work and enables them to focus more on actual teaching.

The total impact of AI on education remains unknown, although it is predicted that in the next 20 years, its influence will increase in this sector (Zawacki-Richter et al. 2019).

8.3.2 The Health Sector Under AI Adoption

In the medical sector, AI is in growing demand. In medicine, AI-based approaches need to be reliable and transparent to perform exceptionally well and be easily explainable to and interpretable by their users (Holzinger et al. 2019). In the field of medicine, AI is proving to be a valuable ally. Deep algorithms are a core solution to managing the growing amounts of data derived from wearable devices, smartphones, and other technologies now commonplace in the medicine. Additionally, AI has brought about a general improvement in the provision of health services by offering more intelligent management of patients' electronic records, assistance with therapeutic compliance, improvement of predictive and preventive medicine, and monitoring of vital functions (Briganti and Le Moine 2020). Additionally, there are improvements in triage, with AI applications such as the Babylon app being used to assist health professionals to identify the patients who need to be examined. The use of AI in the triage of patients has the potential to reduce the burden on health systems and shift the resources to assist those patients who genuinely require medical help (He et al. 2019).

AI can also be deployed for preventive purposes. A significant part of prevention is the promotion of healthy behavior, such as the performance of regular exercise. AI-based health coaches can be used in these scenarios to assist individuals to train others to correctly perform an exercise, help them with their progress, and offer emotional support when they are struggling to succeed. These intelligent coaches also have the advantage of providing personalized support, adapted to the particular need and characteristics of the individual (Mohan et al. 2020).

In the area of mental health, AI is regarded as a valuable solution for some of the current challenges pertaining to treatment and service delivery. Mental health professionals are advocating the use of AI to assist with patient monitoring and to support the decision-making process (Carr 2020). AI-based virtual psychotherapy, for instance, is becoming popular, and it is being delivered by apps that assist individuals to identify patterns of behavior and the mental health conditions associated with them. They help the users to recognize their thoughts and feelings and assist them to develop the skills to address them. These apps, using chatbots, educate the users on relevant clinical terms, and they can provide advice to help the users cope with what they are experiencing. These AI applications are regarded as particularly beneficial for vulnerable and under-served populations (Fiske et al. 2019). Another key aspect of AI in health care is the increasing awareness of its value in low-income countries, as it can address many of the challenges emerging in their health systems. Assisted by the widespread uptake of smartphones and increasing investments in technological development, AI has the potential to improve healthcare delivery in regions that have fewer resources (Wahl et al. 2018).

8.3.3 The Issues and Ethics of AI

Despite abundant evidence of the value of AI to several areas of society, there is a constant dichotomy that is present in all sectors, which is human vs. machines. There is still fear that AI will replace humans, despite current opinions that it will assist, not replace, them (Briganti and Le Moine 2020), since AI has the capacity to complete some tasks with more consistency, speed, and reproducibility than human beings. In medicine, for instance, with the automation of tasks that are theoretically simple, but "incredible labour- and time-intensive, healthcare providers may be freed to tackle more complex tasks, representing an improved use of human capital" (He et al. 2019, p. 30). Another important question pertains to the coexistence of humans and machines and whether AI systems and human beings can coexist effectively. Also, there is the question of which decisions should be made exclusively by humans, which should be delegated to AI, and which decisions should be made collaboratively (Haenlein and Kaplan 2019).

   In the field of medicine, for example, one of the difficulties associated with AI is the absence of a legal framework to establish liability when decisions are driven by algorithms (Briganti and Le Moine 2020). Moreover, the implementation of AI in mainstream medicine raises other issues such as data exchange and privacy, standardization of data, interoperability among different platforms, patient safety and algorithm transparency (He et al. 2019). Algorithm transparency is equally concerning in other fields such as education. Institutions need to exercise caution when granting power to hidden algorithms that can have considerable impact on people's lives but lack transparency. As Popenici and Kerr (2017, p. 4) argues, "this is presented casually as a normal state of facts, the natural arrangements of Internet era, but it translates to highly dangerous levels of unquestioned power." Finally, a popular issue that emerges in regard to AI is its ability to guarantee diversity and reduce bias (Bates et al. 2020). The question of bias emerges in the context of AI foundations, since both the data and the algorithms that are employed for training include the biases that exist in society. Hence, algorithms can be misused and result in breaches of human rights. Different types of biases, such as those related to gender or race can cause inequalities, since they mimic societal stereotypes. The performance of AI algorithms depends on predetermined values, which are subjective and become imprinted in the AI training datasets (Yang et al. 2021).

8.4   IoT and Big Data

IoT can be understood as an "open and comprehensive network of intelligent objects that have the capacity to auto-organize, share information, data and resources, reacting and acting in face of situations and changes in the environment" (Madakam et al. 2015, p. 165). Despite being still in its early stages, IoT is evolving swiftly and becoming one of the most important technologies, with an increase the in number and influence of IoT devices. Its importance is evident in several areas such as transportation, health, smart environments, retail, information technology, and finance (Nord et al. 2019). Big data is different from conventional data due to its three core features: volume, velocity, and variety. Its complex nature demands "powerful technologies and advanced algorithms. So, the traditional static Business Intelligence tools can no longer be efficient" (Oussous et al. 2018, p. 433).

8.4.1   IoT for Smarter Cities

IoT is one of the technologies central to the development of smart cities. The rapid urbanization that is taking place worldwide comes with challenges related to mobility, energy, healthcare provision, and civil infrastructures. Smart cities are intended to be intelligent solutions that guarantee sustainability and efficient use of resources, by embedding intelligent devices in various infrastructures and services. IoT is responsible for connecting the physical and virtual realms by means of widespread devices used in buildings, streets, vehicles, and different infrastructures. It can create integrated solutions for smart cities and offer different types of services to its citizens (Qian et al. 2019). Energy is an integral part of modern cities, and an ongoing concern is the reduction of consumption. The deployment of IoT-enabled devices can assist smart homes to monitor their energy consumption. Mobile phone applications, for example, can be used remotely to manage electric devices (Alavi et al. 2018). The application of IoT for waste management in smart cities is an important advancement in this crucial area. The use of sensors in trucks or garbage cans can provide essential information about the type and amount of garbage and the most appropriate time for collection. This information can also be used to identify the most appropriate locations for placing garbage cans, without detrimental environmental impact or the type of truck that should be used. Insight into these aspects of urban living can help waste management to become more efficient (Samih 2019).

The progress of smart cities is intrinsically connected to big data. "Data in smart cities are characterized by variety, velocity, volume, value, and veracity that are the well-known characteristics of big data." (Karimi et al. 2021, p. 1). In cities, big data can be obtained from individuals and from objects. Data can be collected both from data infrastructures themselves and from energy-consumption behaviors. Moreover, it is possible to gather data on health, public transportation, criminal activities, and the environment (Lim et al. 2018). Big data is essential for supporting various sectors of smart cities; it facilitates traffic management, crime analysis, environmental monitoring, and planning. This big data can be derived from multiple sources including social media, IoT sensors, and different information systems; hence, many factors must be taken into account prior to and during the data analysis. It is important to focus on scalability, interoperability, and data integration and to consider the issues of privacy and security. In addition, it is important to consider streamed data, to enable services that are offered online and in real-time, and historic data for planning and decision-making (Osman 2019). Urban planning, in particular, can benefit significantly from the analysis of big data. Big data allows the mining of real-time data and the detection of high-frequency patterns on a wider scale. In urban planning, big data analytics can make sense of huge volumes of data to empower decisions that in some scenarios can, through automation, reduce the need for human intervention (Kandt and Batty 2021).

8.4.2 E-Government's Adoption of IoT

E-government in another key sector where the IoT is having an important impact. The deployment of IoT can assist administrative management, promote a more intelligent administration, and improve the efficiency of different departments, thereby providing better service to the community. Intelligent e-government platforms based on IoT can be used to support various services and address the challenges deriving from slow servers; moreover, they are more technically advanced than conventional platforms (Qi and Wang 2021). Using real-time data and supported by smart devices, IoT can facilitate knowledge management, information exchange and collaboration not only between different departments and organizations, but also between the citizens and the government. It can assist with decision-making concerning environmental issues and occurrences. For example, it can be used to detect fire in forests and remote areas, monitor the weather, and identify the possibility of extreme natural occurrences such as earthquakes and floods (Papadopoulou et al. 2020). Moreover, the data that results from the communication between all the different devices used by IoT enables local governments to have access to vital information and take timely action as required and offers its citizens more appropriate services and accurate information. IoT can empower governments to improve their public services and at the same time gives citizens the resources allowing them to become more participative (Velsberg et al. 2020).

Big data is essential to IoT in the government sector. Big data can be used to improve the transparency of governments, provide insight to support decision making, offer information about citizens' needs, and help to address pressing problems such as healthcare service delivery, and more sustainable methods for energy production. Moreover, it as a transformational power over procedures and policy-making (Klievink et al. 2017). Big data is even more relevant in the context of e-government where data is more voluminous and derives from an increased electronic participation. Big data offers cost-effective solutions and facilitates the achievement of various e-government objectives, namely in terms of information processing, and knowledge-based decision-making. Big data allows the government to have access to information that was previously unavailable, which offers insight into more aspects of its citizens' lives and behaviors. This allows a more accurate prediction of what the citizens require, enabling a more appropriate response from the government. Furthermore, big data enables a more effective allocation of overall resources (Al-Sai and Abualigah 2017). Besides addressing current issues in e-government, big data can also be used to forecast the future impact of policies. In the context of tax policy, for example, big data can be used to estimate the economic effects of any changes (Agbozo and Spassov 2018).

As Huang and Li (2021, p. 2) posit: "predictions and decisions brought about by big data will inevitably change the way people make decisions. The significance of big data to decision-makers lies in advance prediction, in-process perception, and after-the-fact feedback."

8.4.3 Assessing IoT Challenges

The IoT is not without challenges arising from issues such as the integration of innovative technology with existing technology, privacy concerns, security issues and questions of trust (Nord et al. 2019). One of the main challenges of IoT is security, as cyberattacks are a risk to IoT devices and to entire systems, such as health care. Another aspect that needs to be considered is the environmental impact of the increasing uptake of devices and the amount of energy that is required for their operation. Finally, more effort should be made to design and develop software that is user-friendly (Nižetić et al. 2020). Within e-government, for example, at a technical level, the diversity of IoT devices can cause difficulties in terms of interoperability and communication between them, and problems of security given that diverse devices are being used to access governmental services. Furthermore, there are other challenges related to: the required infrastructure such as sensors, digital tags and storage and data management solutions to address the rising volume of data; and the incorporation of AI, which may enable devices and systems to make decisions. At a non-technical level, it is important to: know the reasons for individuals' and organizations' reluctance to adopt IoT; consider the required organizational reform and the redesign of certain processes; allocate the necessary financial resources to enable such changes; establish an institutional and legal framework; to reflect on security, privacy, and trust; and invest in collaborative partnerships with other entities, particularly in the private sector (Papadopoulou et al. 2020).

There are also various difficulties associated with big data, one of which concerns the management of data quality. Different sources of data can produce redundant or contradicting information, which reduces the accuracy of the insights that are obtained which, subsequently, can result in poor decisions. Standardization and the integration of data from a variety of sources are difficult tasks when dealing with big data and require expertise and a significant amount of time (Lim et al. 2018).

Furthermore, big data deals with different types of data, some of which can be imprecise. This is the case with sentiment data which is gathered via social media and is subjective and relies on human judgment and therefore devoid of objectivity. This subjectivity becomes a challenge, when it comes to, for example, machine learning, because it is difficult to learn from this type of data (L'heureux et al. 2017). Big data also poses challenges in terms of managing privacy and security, data governance, and the exchange of information and data. Also, at this management level, it is important to consider the intricacies related to data ownership and to the financial investment it requires. Even though big data can be a valuable business ally, it requires interpretation and analysis, which demand particular skills that are not always readily available (Sivarajah et al. 2017).

8.5   XR and the Boundaries Between Realities

XR can be understood as an "umbrella term to represent all computer-mediated reality technologies that merge the physical and virtual worlds for the enhanced experience." (Gong et al. 2021), "encapsulating augmented reality (AR), virtual reality (VR), and mixed reality (MR)" (Chuah 2018, p. 1).

8.5.1   XR in Healthcare

AR and VR seem to be more popular in the health sector, especially in the field of surgery, for training and assisting performance; in psychology to assist patients with certain conditions, such as phobias; and in rehabilitation to help victims of stroke, for example (Muñoz-Saavedra et al. 2020). Surgery can definitely benefit from XR. In a study of cardiothoracic surgery, it was concluded that, although the use of XR for surgical is still in its infancy, it can be a valuable means of improving preoperative planning, using virtual reality. In this study (Sadeghi et al. 2020), intraoperative navigation, combined with reality and augmented reality, guided the surgical procedure. Moreover, a study on the use of AR in orthopedic surgery reported that its use is increasing and attracting more interest; the benefits include better accuracy while performing surgery, less exposure to radiation, and reduction of surgery time (Casari et al. 2021).

In the mental health arena, virtual reality can be used to create outdoor experiences, especially for those without access to the natural environment, and promotes health and emotional well-being (Browning et al. 2020). In addition, VR can be used to offer exposure-based techniques to address mental health diseases. The patients have the opportunity to, safely, confront their phobias within a controlled environment. This type of treatment has proven to be effective for numerous conditions. VR is an important tool for clinical assessment, where real-world experiences can be simulated to allow the clinician to observe the patients in a situation that mimics what occurs in their daily lives (e.g., anxiety, paranoia, fear). The usefulness of VR in the mental health domain is not limited to treatment or clinical assessment; it is also a valuable research tool. It can, for example, be used to conduct research on dangerous or inaccessible scenarios, and in neuroimaging research, to explore brain activation within naturalistic contexts (Bell et al. 2020).

In rehabilitation, VR and AR technologies are showing their potential by offering novel experiences to patients during their rehabilitation sessions and, consequently, strengthening their engagement. This increased patient engagement can improve their physical outcomes. Moreover, these technologies enable the creation of "remote physical therapy," where AR and VR are implemented in a IoT ecosystem to collect, store, and analyze data on the performance of the patient, enabling the physiotherapist to evaluate progress and outcomes in order to design and implement tailored recovery plans (Postolache et al. 2021). In the context of post-stroke rehabilitation, the benefits of VR have been acknowledged by practitioners as they can increase therapy time and offer supplementary sessions. In addition, generally speaking, VR can improve patient motivation in the sense that individuals are willing to practice rehabilitation exercises more frequently and/or with more intensity as a result of the engaging scenarios (Levin 2020).

8.5.2   The Business Affordances of XR

Immersive technologies offer to consumers the opportunity to enhance their engagement and become co-creators with companies. XR makes it possible to improve the experience of the customer in many areas. When visiting an art gallery, for example, visitors are able to scan the brochure for more information using an AR application; they can, in a virtual environment, engage in a game with historical avatar. During the visit, it is equally possible to use AR to obtain more information about the displayed art of the pieces of art and take advantage of VR to be transported to a remote setting (Flavián et al. 2019).

XR technology has the potential to assist with a wide range of manufacturing activities, including all processes from design to service. The traditional work routines in manufacturing companies can be greatly improved by XR systems (Gong et al. 2021). In regard to the automotive industry, VR can be used in product design to improve vehicle safety. For example, it can be used to help assess the drivers' visibility and, in terms of ergonomics, to establish the best location for a door handle or to determine the reachability of the control panel. This assessment can be performed with VR, and several features can be rearranged in the virtual cabin (Berg and Vance 2017). Also, within the industry sector, there is growing potential for AR to improve technical manuals, which previous research has shown to be clearer than other formats (Gattullo et al. 2019).

Augmented reality (AR) is one of the most promising technologies for technical manuals in the context of Industry 4.0. However, the implementation of AR documentation in industry is still challenging because specific standards and guidelines have yet to be established. In this work, we propose a novel methodology for the conversion of existing "traditional" documentation, and for the authoring of new manuals in AR in compliance with Industry 4.0 principles. The methodology is based on the optimization of text usage with the ASD Simplified Technical English, the conversion of text instructions into 2D graphic symbols, and the structuring of the content through the combination of Darwin Information Typing Architecture (DITA) and Information Mapping (IM). We tested the proposed approach with a case study of a maintenance manual for hydraulic brakes. We validated it with a user test followed by the collection of subjective feedback from 22 users. The results of this experiment confirm that the manual produced by our methodology is clearer than other templates.

In real estate, there are AR applications that enable house hunters to scan neighborhoods for possible houses, providing them with information on sale prices, taxes, the actual size of the lot, and other details allowing prospective buyers to make informed decisions. When shopping for groceries, clients with dietary constraints can examine products with AR-enabled smart glasses and identify whether the product is safe or according to their specific needs. VR can offer 360-degree tours of remote places, and it can be a powerful tool for businesses with online shopping as it provides more accurate depictions of their products (Farshid et al. 2018). Additionally, in retail, AR allows consumers to see the products, and in some cases, virtually try them, prior to purchasing (Slater et al. 2020).

8.5.3   The Challenges of Bridging Realities

Despite the fact that AR and VR applications are becoming increasingly user-friendly and accessible, there are several difficulties in their creation. The creators of AR and VR applications face several barriers such as difficulty with specific guidelines and examples of design and rely on of story-driven experiences when designing a product (Ashtari et al. 2020). There is also the issue of cost, which is high in terms of equipment since high-end systems are required to run VR environments. The development of VR applications is financially demanding, as is their maintenance and associated devices (Garrett et al. 2018). In addition, XR systems face substantial challenges concerning data exchange between different systems. Data incompatibility affects the quality of visualization. This is an important issue for these systems as they are subjected to heavy data interchange. Moreover, they present some issues in terms of interactive design (Gong et al. 2021). In the context of mixed-reality (MR) devices, certain scenarios could present significant challenges. For instance, people who use MR headsets while traveling could experience motion sickness, hinder the effectiveness of safety mechanisms, such as seat belts and airbags in case of a car accident, and they can display socially reprehensible behavior (McGill et al. 2020). In addition, since MR platforms offer interaction between reality and virtuality, they demand the deployment of precise tracking methods for both types of objects, which can be challenging. Another difficulty arising from MR platforms pertains to the need for suitable display technology capable of ensuring adequate resolution as well as contrast (Rokhsaritalemi et al. 2020).

In regard to AR, when used in education, despite various reported benefits (Isaias and Reis 2016), it has been associated with cognitive overload, an excessive amount of information, high cost of implementation and technical complexity (Alzahrani 2020). Furthermore, when users are presented with AR, they tend to overly concentrate on the content that is being presented virtually, while being oblivious to their immediate physical environment. In situations where AR is used together with other formats, and information is being delivered by humans as well, it can distract users and cause them to miss important information, undermining the role of the human instructor (Syiem et al. 2020). Regardless of all the benefits associated with VR, it is essential to understand that VR technology is valuable, not as a replacement for the real experience, but as a complementary tool that can assist individuals to engage in and benefit from real-world experiences (Flavián et al. 2019). Also, VR systems require users to have some technical expertise; otherwise, users may experience cybersickness, headaches, or eye strain and the head-mounted devices used for VR can be uncomfortable (Garrett et al. 2018). Finally, it is important to consider the implications of using systems that mimic reality so closely. The evolution of VR has improved the experience of the user in terms of immersion, with virtual worlds and characters becoming increasingly similar to reality. When immersed in these scenarios, which blur the boundary between reality and virtuality, users can become confused about what is real and what is not (Pan and Hamilton 2018).

8.6   Reflecting About the Future of ICTs

The aforementioned trends and technologies suggest a positive future where technologies, despite their shortcomings, can improve and advance society. Nonetheless, as technology progresses and introduces into society innovations that are applicable in all sectors, it is important to examine two fundamental aspects of increased digitalization and rapid technological development: digital divide and sustainability.

8.6.1   Sustainability

Sustainability is a concept that is becoming increasingly more important in an era when there is increased awareness of planet Earth's urgent need for solutions to its ecological, economic, and social issues. Hence, sustainability encompasses a range of interconnected problems concerning these three dimensions: endangered ecosystems, poverty, and resource depletion. The pursuit of sustainability is an attempt to foster systems and processes with the ability to survive in the long-term future and to ensure a sustainable ecology, promote economic opportunity and foster social inclusion (Robertson 2021).

At an environmental level, ICTs have the potential to reduce CO_2 emissions, particularly since they play a pivotal role in conserving energy in smart cities, improving transportation and electrical grids and developing smarter industries and energy-saving solutions. Paradoxically, ICTs also have a detrimental effect on the environment: they require a great deal of technological equipment and devices; they consume enormous amounts of energy; and they have huge impact on electronic waste recycling. For this relationship to become positive, ICTs need to reach, as is already happening in some developed countries, highly sophisticated levels of development (Añón Higón et al. 2017). In order to minimise the negative impact of digitalization, companies must focus on developing sustainable digital solutions that can advance their digitalization and sustainability efforts simultaneously (Lichtenthaler 2021). The term "digitainability" refers exactly to this relationship, as it concerns the "cross-fertilisation between the process of digitalisation and sustainable development" (Gupta et al. 2020, p. 9285).

Concerning the economy, ICTs have the potential to foster economic growth. However, at the same time, they can exclude certain populations, thereby exacerbating economic disparities. Thus, it is important to ensure a more equitable access to and use of ICTs. In terms of what ICTs can specifically do for economic prosperity, the possibilities include the use of big data to obtain information on those who live in poverty. The data mined from sensors and social media can inform interventions; increased access to mobile phones can assist new entrepreneurs to establish their businesses; and online learning platforms can facilitate training and knowledge transfer (International Telecommunication Union 2017). In agriculture, ICTs and big data can assist farmers with the timely identification of epidemics and plagues (Tjoa and Tjoa 2016).

ICTs can play an important role in the promotion of social equity and well-being. In the health sector, ICTs are revolutionizing the delivery of healthcare services by providing digital solutions, such as telemedicine at home, to alleviate the pressure on hospitals; offering online educational material to medical professionals worldwide; and deploying a variety of technologies that assist in the monitoring of health conditions. ICTs play an important role in promoting social equality, and improving communication through mobile phones and Internet connectivity (International Telecommunication Union 2017). Moreover, ICTs can be used in the education domain to give access to education to more people globally, through the development of online platforms for learning, following the MOOC format (Tjoa and Tjoa 2016).

8.6.2 Digital Divide

It is estimated that around 53% of the world population has access to the Internet; hence, a considerable number of people still have little or no possibility of accessing online platforms and reaping their benefits (United Nations Conference on Trade and Development 2020). In an era marked by rapidly evolving ICTs and increasing digitalization, those who have no access to technologies because increasingly distanced from those who have the means to access and master technology. Hence, it is important to reflect on all aspects of the digital divide.

The notion of digital divide has evolved from being a term focused on access, to a more encompassing conceptualization that includes other inequalities regarding the use of the Internet. The digital divide has three levels: level one pertains to Internet access itself; level two refers to the disparities that emerge from the individuals' own motivations, the skills that they have, and the purpose(s) for which they use the Internet; level three concerns the disparities related to the various social, economic, personal, political benefits that the individuals can obtain when accessing the Internet (Ragnedda and Ruiu 2017). The digital divide has a range of consequences that have impacts on most areas of life and daily living. The increasing digitalization of society has caused, in the majority of developed countries, an expectation that everyone has access to the Internet, more specifically, when communicating with the government, when applying for a job, for socializing with friends, and for entertainment. Hence, inaccessibility can exclude people economically, socially, and culturally. Moreover, globally speaking, the lack of access can have a detrimental effect on innovation, development, and economic prosperity (Van Dijk 2020). Overall, the digital divide results in limited opportunities for those who are already disadvantaged. The effects of the digital divide have become even more apparent with the COVID-19 pandemic, with people having to use the Internet to work and/or continue their education, and for social support and information. The lack of access to the Internet has compromised the quality of people's day-to-day lives during lockdown periods (Lai and Widmar 2020).

Addressing the digital divide requires international cooperation and an investment in key areas. It is essential that ICT infrastructures be developed or improved, particularly in poorly served areas where access is very restricted. Apart from being accessible, the Internet needs to be efficient, quick, and reliable. In addition, individuals' literacy, numeracy, and digital skills need to be improved, with a particular consideration for marginalized populations, senior citizens, women, and individuals with disabilities. Hence, it is important that the development of skills be inclusive (United Nations Conference on Trade and Development 2021). One of the most significant difficulties of connecting under-served areas is the financial cost that it incurs and these areas' low income. To address these issues, 6G can be used to improve both the heterogeneity and the scale of the network, at the same time improving the performance of the network in general (Chaoub et al. 2021).

8.7 Conclusion

The increasing digitalization of society is welcomed in many sectors, which currently benefit from all the advantages it brings. Nonetheless, despite the increasing pervasiveness of technology, many organizations and individuals are reluctant to adopt technological innovations that are perceived to have disadvantages or shortcomings.

This chapter examined several innovative technologies, focusing on their advantages as well as on the difficulties associated with them.

The role of social networks in education was examined, particularly in regard to their potential in building online learning communities and strengthening online interaction. In the business and industrial sectors, ICTs can streamline operations and procedures and provide opportunities for career advancement and insight into customers' preferences. In regard to challenges, two essential themes were discussed: anonymity/privacy and the dissemination of misinformation/disinformation. The role of AI in education was investigated, including its benefits ranging from personalization to improved learning/teaching performance. In the healthcare sector, AI can offer intelligent coaches and automated solutions. Ethical questions and challenges were related to the possible replacement of humans and algorithm transparency. IoT and big data were explored in the context of smart cities for their support of data-driven decisions and intelligent solutions and in e-government for their role in providing vital information and improving service delivery. The issues that were identified pertained, for instance, to interoperability and data quality. The advantages of XR within the healthcare sector were presented, both for physical recovery and for mental health, and within the business sector, for an improved customer experience and more efficient industrial processes. The difficulties associated with XR relate to issues such as technical complexity and a blurring between reality and virtuality.

Looking forward, it is important to ensure the sustainability of these technologies and a more equal distribution of access and skills, which will help to bridge the digital divide. Future research could complement this conceptual account of these technologies incorporating the views of the experts in the different areas, using focus groups or semi-structured interviews.

References

Agbozo E, Spassov K (2018) Establishing efficient governance through data-driven e-government. In: Proceedings of the 11th international conference on theory and practice of electronic governance, pp 662–664

Ahmad T, Zhang D, Huang C, Zhang H, Dai N, Song Y, Chen H (2021) Artificial intelligence in sustainable energy industry: status Quo, challenges, and opportunities. J Clean Prod 289:125834. https://doi.org/10.1016/j.jclepro.2021.125834

Al-Sai ZA, Abualigah LM (2017) Big data, and E-government: a review. In: 2017 8th international conference on information technology (ICIT). IEEE, pp 580–587

Alavi AH, Jiao P, Buttlar WG, Lajnef N (2018) Internet of things-enabled smart cities: state-of-the-art and future trends. Measurement 129:589–606

Alshaikh F, Ramzan F, Rawaf S, Majeed A (2014) Social network sites as a mode to collect health data: a systematic review. J Med Internet Res 16(7):e171

Alzahrani NM (2020) Augmented reality: a systematic review of its benefits and challenges in e-learning contexts. Appl Sci 10(16):5660

Amoruso M, Anello D, Auletta V, Cerulli R, Ferraioli D, Raiconi A (2020) Contrasting the spread of misinformation in online social networks. J Artif Intell Res 69:847–879

Añón Higón D, Gholami R, Shirazi F (2017) ICT and environmental sustainability: a global perspective. Telemat Inform 34(4):85–95. https://doi.org/10.1016/j.tele.2017.01.001

Ashtari N, Bunt A, McGrenere J, Nebeling M, Chilana PK (2020) Creating augmented and virtual reality applications: current practices, challenges, and opportunities. In: Proceedings of the 2020 CHI conference on human factors in computing systems, pp 1–13

Bastick Z (2021) Would you notice if fake news changed your behavior? An experiment on the unconscious effects of disinformation. Comput Human Behav 116:106633

Bates T, Cobo C, Mariño O, Wheeler S (2020) Can artificial intelligence transform higher education? SpringerOpen

Bell IH, Nicholas J, Alvarez-Jimenez M, Thompson A, Valmaggia L (2020) Virtual reality as a clinical tool in mental health research and practice. Dialogues Clin Neurosci 22(2):169

Benson V, Morgan S, Filippaios F (2014) Social career management: social media and employability skills gap. Comput Hum Behav 30:519–525

Berg LP, Vance JM (2017) Industry use of virtual reality in product design and manufacturing: a survey. Virtual Real 21(1):1–17

Briganti G, Le Moine O (2020) Artificial intelligence in medicine: today and tomorrow. Front Med 7:27

Browning MHEM, Mimnaugh KJ, van Riper CJ, Laurent HK, LaValle SM (2020) Can simulated nature support mental health? Comparing short, single-doses of 360-degree nature videos in virtual reality with the outdoors. Front Psychol 10(2667). https://doi.org/10.3389/fpsyg.2019.02667

Buccafurri F, De Angelis V, Idone MF, Labrini C (2021) Enabling location k-anonymity in social networks. In: Securing social networks in cyberspace. CRC Press, pp 35–50

Carr S (2020) 'AI gone mental': engagement and ethics in data-driven technology for mental health. Taylor & Francis

Casari FA, Navab N, Hruby LA, Kriechling P, Nakamura R, Tori R, dos Santos Nunes FdL, Queiroz MC, Fürnstahl P, Farshad M (2021) Augmented reality in orthopedic surgery is emerging from proof of concept towards clinical studies: a literature review explaining the technology and current state of the art. Curr Rev Musculoskelet Med 1–12

Chaoub A, Giordani M, Lall B, Bhatia V, Kliks A, Mendes L, Rabie K, Saarnisaari H, Singhal A, Zhang N (2021) 6G for bridging the digital divide: wireless connectivity to remote areas. IEEE Wirel Commun

Chen L, Chen P, Lin Z (2020) Artificial intelligence in education: a review. IEEE Access 8:75264–75278

Chuah SH-W (2018) Why and who will adopt extended reality technology? Literature review, synthesis, and future research agenda

Clark J-A, Liu Y, Isaias P (2020) Critical success factors for implementing learning analytics in higher education: a mixed-method inquiry. Australas J Educ Technol 36(6):89–106

Cruz-Benito J, Borrás-Gené O, García-Peñalvo FJ, Blanco ÁF, Therón R (2017) Learning communities in social networks and their relationship with the MOOCs. IEEE Rev Iberoam Tecnol Aprendiz 12(1):24–36. https://doi.org/10.1109/RITA.2017.2655218

Davis J, Wolff H-G, Forret ML, Sullivan SE (2020) Networking via LinkedIn: an examination of usage and career benefits. J Vocat Behav 118:103396. https://doi.org/10.1016/j.jvb.2020.103396

Falcão J, Isaías P (2020) Perceptions and attitude toward advertising on social networks and search engines: a comparative analysis. J Internet Commer 19(4):404–436. https://doi.org/10.1080/15332861.2020.1816325

Farshid M, Paschen J, Eriksson T, Kietzmann J (2018) Go boldly! Explore augmented reality (AR), virtual reality (VR), and mixed reality (MR) for business. Bus Horiz 61(5):657–663

Fiske A, Henningsen P, Buyx A (2019) Your robot therapist will see you now: ethical implications of embodied artificial intelligence in psychiatry, psychology, and psychotherapy. J Med Internet Res 21(5):e13216

Flavián C, Ibáñez-Sánchez S, Orús C (2019) The impact of virtual, augmented, and mixed reality technologies on the customer experience. J Bus Res 100:547–560. https://doi.org/10.1016/j.jbusres.2018.10.050

Garrett B, Taverner T, Gromala D, Tao G, Cordingley E, Sun C (2018) Virtual reality clinical research: promises and challenges. JMIR Serious Games 6(4):e10839

Gattullo M, Scurati GW, Fiorentino M, Uva AE, Ferrise F, Bordegoni M (2019) Towards augmented reality manuals for industry 4.0: a methodology. Robot Comput Integr Manuf 56:276–286. https://doi.org/10.1016/j.rcim.2018.10.001

Giurgiu L (2017) Microlearning an evolving elearning trend. Sci Bull Nicolae Balcescu Land Forces Acad 22(1):18–23

Gong L, Fast-Berglund Å, Johansson B (2021) A framework for extended reality system development in manufacturing. IEEE Access 9:24796–24813

Gonzalez R, Llopis J, Gasco J (2015) Social networks in cultural industries. J Bus Res 68(4):823–828

Gupta S, Motlagh M, Rhyner J (2020) The digitalization sustainability matrix: a participatory research tool for investigating digitainability. Sustainability 12(21):9283

Haenlein M, Kaplan A (2019) A brief history of artificial intelligence: on the past, present, and future of artificial intelligence. Calif Manage Rev 61(4):5–14

He J, Baxter SL, Xu J, Xu J, Zhou X, Zhang K (2019) The practical implementation of artificial intelligence technologies in medicine. Nat Med 25(1):30–36

Holzinger A, Langs G, Denk H, Zatloukal K, Müller H (2019) Causability and explainability of artificial intelligence in medicine. Interdiscip Rev Data Mining Knowl Discov (Wiley) 9(4):e1312

Hortigüela-Alcalá D, Sánchez-Santamaría J, Pérez-Pueyo Á, Abella-García V (2019) Social networks to promote motivation and learning in higher education from the students' perspective. Innov Educ Teach Int 56(4):412–422. https://doi.org/10.1080/14703297.2019.1579665

Huang M-H, Rust RT (2021) A strategic framework for artificial intelligence in marketing. J Acad Mark Sci 49(1):30–50

Huang Q, Li XW (2021) Research on the design of government affairs platform in the context of big data. Sci Program 2021:9936217. https://doi.org/10.1155/2021/9936217

Hudson S, Roth MS, Madden TJ, Hudson R (2015) The effects of social media on emotions, brand relationship quality, and word of mouth: an empirical study of music festival attendees. Tour Manage 47:68–76

Imran AS, Pireva K, Dalipi F, Kastrati Z (2016) An analysis of social collaboration and networking tools in e-learning. In: 2016 learning and collaboration technologies. Springer International Publishing, Cham, pp 332–343

International Telecommunication Union (2017) Fast-forward progress: leveraging tech to achieve the global goals. United Nations Publications, NY, USA

Isaías P (2018) Model for the enhancement of learning in higher education through the deployment of emerging technologies. J Inf Commun Ethics Soc 16(4):401–412. https://doi.org/10.1108/JICES-04-2018-0036

Isaías P, Pífano S, Miranda P (2012) Social network sites: modeling the new business-customer relationship. In: Social networking and community behavior modeling: qualitative and quantitative measures. IGI Global, pp 248–265. https://doi.org/10.4018/978-1-61350-444-4.ch014

Isaias P, Reis F (2016) Empathic technologies and virtual, contextual, and mobile learning in VR/AR/MR environments. Proc Int Mobile Learn Festiv 3–12

Isaias P, Reis F, Coutinho C, Lencastre JA (2017) Empathic technologies for distance/mobile learning. Interact Technol Smart Educ 14(2):159–180. https://doi.org/10.1108/ITSE-02-2017-0014

Kandt J, Batty M (2021) Smart cities, big data, and urban policy: towards urban analytics for the long run. Cities 109:102992. https://doi.org/10.1016/j.cities.2020.102992

Kaplan A, Haenlein M (2019) Siri, Siri, in my hand: who's the fairest in the land? On the interpretations, illustrations, and implications of artificial intelligence. Bus Horiz 62(1):15–25

Karimi Y, Haghi Kashani M, Akbari M, Mahdipour (2021) E leveraging big data in smart cities: a systematic review. Concurr Comput Pract Exp e6379. https://doi.org/10.1002/cpe.6379

Kataoka H, Ogawa Y, Echizen I, Kuboyama T, Yoshiura H (2014) Effects of external information on anonymity and role of transparency with example of social network de-anonymisation. In: 2014 ninth international conference on availability, reliability and security (ARES), 8–12, 2014, pp 461–467. https://doi.org/10.1109/ARES.2014.70

Klein A, Ahlf H, Sharma V (2015) Social activity and structural centrality in online social networks. Telemat Inform 32(2):321–332

Klievink B, Romijn B-J, Cunningham S, de Bruijn H (2017) Big data in the public sector: uncertainties and readiness. Inf Syst Front 19(2):267–283

L'heureux A, Grolinger K, Elyamany HF, Capretz MA (2017) Machine learning with big data: challenges and approaches. IEEE Access 5:7776–7797

Lachman R, Joffe M (2021) Applications of artificial intelligence in media and entertainment. In: Thomas Heinrich M, Adrian David C (eds) Analyzing future applications of AI, sensors, and robotics in society. IGI Global, Hershey, PA, USA, pp 201–220. https://doi.org/10.4018/978-1-7998-3499-1.ch012

Lai J, Widmar NO (2020) Revisiting the digital divide in the COVID-19 era. Appl Econ Perspect Policy. https://doi.org/10.1002/aepp.13104.10.1002/aepp.13104

Levin MF (2020) What is the potential of virtual reality for post-stroke sensorimotor rehabilitation? Expert Rev Neurother 20(3):195–197. https://doi.org/10.1080/14737175.2020.1727741

Lichtenthaler U (2021) Digitainability: the combined effects of the megatrends digitalization and sustainability. J Innov Manag 9(2):64–80

Lim C, Kim K-J, Maglio PP (2018) Smart cities with big data: reference models, challenges, and considerations. Cities 82:86–99. https://doi.org/10.1016/j.cities.2018.04.011

Madakam S, Lake V, Lake V, Lake V (2015) Internet of things (IoT): a literature review. J Comput Commun 3(05):164

McGill M, Williamson J, Ng A, Pollick F, Brewster S (2020) Challenges in passenger use of mixed reality headsets in cars and other transportation. Virtual Real 24(4):583–603

Miranda P, Isaias P, Costa CJ, Pifano S (2017) Validation of an e-learning 3.0 critical success factors framework: a qualitative research. J Inf Technol Educ Res 16(1):339–363

Mohan S, Venkatakrishnan A, Hartzler AL (2020) Designing an AI health coach and studying its utility in promoting regular aerobic exercise. ACM Trans Interact Intell Syst (TiiS) 10(2):1–30

Muñoz-Saavedra L, Miró-Amarante L, Domínguez-Morales M (2020) Augmented and virtual reality evolution and future tendency. Appl Sci 10(1):322

Ng TK (2021) New interpretation of extracurricular activities via social networking sites: a case study of artificial intelligence learning at a secondary school in Hong Kong. J Educ Train Stud 9(1):49–60

Nižetić S, Šolić P, González-de DL-d-I, Patrono L (2020) Internet of things (IoT): opportunities, issues, and challenges towards a smart and sustainable future. J Clean Prod 274:122877

Nord JH, Koohang A, Paliszkiewicz J (2019) The internet of things: review and theoretical framework. Expert Syst Appl 133:97–108

Olvera-Lobo M-D (2020) Websites and social networks. A study of healthcare SMEs in Andalusia. In: Information technology and systems: proceedings of ICITS 2020 1137, p 297

Osman AMS (2019) A novel big data analytics framework for smart cities. Futur Gener Comput Syst 91:620–633

Oussous A, Benjelloun F-Z, Ait Lahcen A, Belfkih S (2018) Big data technologies: a survey. J King Saud Univ Comput Inf Sci 30(4):431–448. https://doi.org/10.1016/j.jksuci.2017.06.001

Pan X, Hamilton AFdC (2018) Why and how to use virtual reality to study human social interaction: the challenges of exploring a new research landscape. Br J Psychol 109(3):395–417

Papadopoulou P, Kolomvatsos K, Hadjiefthymiades S (2020) Internet of things in E-government: applications and challenges. Int J Artif Intell Mach Learn (IJAIML) 10(2):99–118

Peddinti ST, Ross KW, Cappos J (2014) On the internet, nobody knows you're a dog: a twitter case study of anonymity in social networks. In: Proceedings of the second edition of the ACM conference on online social networks. ACM, pp 83–94

Peng S, Zhou Y, Cao L, Yu S, Niu J, Jia W (2018) Influence analysis in social networks: a survey. J Netw Comput Appl 106:17–32

Persia F, Auria DD (2017) A survey of online social networks: challenges and opportunities. In: 2017 IEEE international conference on information reuse and integration (IRI), 4–6, pp 614–620. https://doi.org/10.1109/IRI.2017.74

Pifano S, Isaias P, Miranda P (2021) Successful implementation of web 2.0 in non-profit organisations: a case study. In: Handbook of research on user experience in web 2.0 technologies and its impact on universities and businesses. IGI Global, pp 353–369

Popenici SAD, Kerr S (2017) Exploring the impact of artificial intelligence on teaching and learning in higher education. Res Pract Technol Enhanc Learn 12(1):22. https://doi.org/10.1186/s41039-017-0062-8

Postolache O, Monge J, Alexandre R, Geman O, Jin Y, Postolache G (2021) Virtual reality and augmented reality technologies for smart physical rehabilitation. In: Kanoun O, Derbel N (eds) Advanced systems for biomedical applications. Springer International Publishing, Cham, pp 155–180. https://doi.org/10.1007/978-3-030-71221-1_8

Qi M, Wang J (2021) Using the internet of things E-government platform to optimize the administrative management mode. Wirel Commun Mobile Comput

Qian Y, Wu D, Bao W, Lorenz P (2019) The internet of things for smart cities: technologies and applications. IEEE Netw 33(2):4–5

Quadir B, Chen N-S, Isaias P (2020) Analyzing the educational goals, problems and techniques used in educational big data research from 2010 to 2018. Interact Learn Environ 1–17

Quimbita A, Pupiales A, Guerrero G (2020) Proposal to improve the usability of social networks using eye tracking: a study to optimize internal communication in the university context. In: 2020 15th Iberian conference on information systems and technologies (CISTI), 24–27, pp 1–6. https://doi.org/10.23919/CISTI49556.2020.9140883

Ragnedda M, Ruiu M (2017) Social capital and the three levels of digital divide

Robertson M (2021) Sustainability principles and practice, 3rd edn. Routledge, Oxon, UK/NY, NY, USA

Rokhsaritalemi S, Sadeghi-Niaraki A, Choi S-M (2020) A review on mixed reality: current trends, challenges, and prospects. Appl Sci 10(2):636

Rosabel M-R, Juan-Francisco Á-H, Mayra U-S (2022) Social networks and blogs as educational communicative resources during a pandemic. In: Antonio Cortijo O, Vicent M (eds) Handbook of research on historical pandemic analysis and the social implications of COVID-19. IGI Global, Hershey, PA, USA, pp 153–163. https://doi.org/10.4018/978-1-7998-7987-9.ch012

Sadeghi AH, El Mathari S, Abjigitova D, Maat AP, Taverne YJ, Bogers AJ, Mahtab EA (2020) Current and future applications of virtual, augmented, and mixed reality in cardiothoracic surgery. Ann Thorac Surg

Samih H (2019) Smart cities and internet of things. J Inf Technol Case Appl Res 21(1):3–12

Shelomovska O, Romaniukha M, Sorokina N (2017) Social networks as a constituent of e-learning environment. Sciendo 6(4):4–14

Shestak V, Gura A, Borisova U, Kozlovskaya D (2021) The role of social networks in the organization of the educational process and learning. Int J Interact Mobile Technol 15(11)

Sitaula N, Mohan CK, Grygiel J, Zhou X, Zafarani R (2020) Credibility-based fake news detection. In: Disinformation, misinformation, and fake news in social media. Springer, pp 163–182

Sivarajah U, Kamal MM, Irani Z, Weerakkody V (2017) Critical analysis of big data challenges and analytical methods. J Bus Res 70:263–286. https://doi.org/10.1016/j.jbusres.2016.08.001

Slater M, Gonzalez-Liencres C, Haggard P, Vinkers C, Gregory-Clarke R, Jelley S, Watson Z, Breen G, Schwarz R, Steptoe W, Szostak D, Halan S, Fox D, Silver J (2020) The ethics of realism in virtual and augmented reality. Front Virtual Real 1(1). https://doi.org/10.3389/frvir.2020.00001

Sun Z, Anbarasan M, Praveen Kumar D (2021) Design of online intelligent English teaching platform based on artificial intelligence techniques. Comput Intell 37(3):1166–1180. https://doi.org/10.1111/coin.12351

Syiem BV, Kelly RM, Velloso E, Goncalves J, Dingler T (2020) Enhancing visitor experience or hindering docent roles: attentional issues in augmented reality supported installations. In: 2020 IEEE international symposium on mixed and augmented reality (ISMAR). IEEE, pp 279–288

Teoh K-K, Pourshafie T, Balakrishnan V (2014) A gender lens perspective of the use of social network in higher education in Malaysia and Australia. In: Proceedings of the 2014 international conference on social computing. ACM, p 21

Tjoa AM, Tjoa S (2016) The role of ICT to achieve the UN sustainable development goals (SDG). In: ICT for promoting human development and protecting the environment. Springer International Publishing, Cham, pp 3–13

United Nations Conference on Trade and Development (2020) Fifteen years since the world summit on the information society. United Nations Publications, New York, NY, USA

United Nations Conference on Trade and Development (2021) Technology and innovation report 2021: catching technological waves innovation with equity. United Nations Publications, New York, NY, USA

Van Dijk J (2020) The digital divide. Wiley

Vanderhoven E, Schellens T, Valcke M (2014) Educational packages about the risks on social network sites: state of the art. Procedia Soc Behav Sci 112:603–612

Velsberg O, Westergren UH, Jonsson K (2020) Exploring smartness in public sector innovation-creating smart public services with the internet of things. Eur J Inf Syst 29(4):350–368

Wahl B, Cossy-Gantner A, Germann S, Schwalbe NR (2018) Artificial intelligence (AI) and global health: how can AI contribute to health in resource-poor settings? BMJ Glob Health 3(4):e000798

Yang SJH, Ogata H, Matsui T, Chen N-S (2021) Human-centered artificial intelligence in education: seeing the invisible through the visible. Comput Educ Artif Intell 2:100008. https://doi.org/10.1016/j.caeai.2021.100008

Zawacki-Richter O, Marín VI, Bond M, Gouverneur F (2019) Systematic review of research on artificial intelligence applications in higher education–where are the educators? Int J Educ Technol High Educ 16(1):1–27

Zhang P, Guo Z, Ullah S, Melagraki G, Afantitis A, Lynch I (2021) Nanotechnology and artificial intelligence to enable sustainable and precision agriculture. Nat Plants 7(7):864–876. https://doi.org/10.1038/s41477-021-00946-6

Zhou X, Zafarani R, Shu K, Liu H (2019) Fake news: fundamental theories, detection strategies and challenges. In: Proceedings of the twelfth ACM international conference on web search and data mining, Melbourne VIC, Australia. Association for Computing Machinery, New York, NY, USA, pp 836–837. https://doi.org/10.1145/3289600.3291382

Index

© Springer-Verlag London Ltd., part of Springer Nature 2022
T. Issa and P. Isaias, *Sustainable Design*,
https://doi.org/10.1007/978-1-4471-7513-1

Printed in the United States
by Baker & Taylor Publisher Services